THE FUNDAMENTALS OF DRONE PHOTOGRAMMETRY, MAPPING, AND SURVEY

By

Thomas Dowell

Contents

Drone Photogrammetry, and Mapping, Survey

Preface

This book is designed to be both a text book and a reference guide, as the author, I would encourage you to use it as both. For that reason, this book is consistently redundant. You will find similar references throughout the text and chapters herein. The goal is to provide a complete picture of the source material. The philosophy being, it is better to have too much information than not enough.

Welcome to "The Fundamentals of Drone Photogrammetry, Mapping, and Survey" a comprehensive guide designed for drone pilots and inspection, mapping, and survey professionals entering the field of geospatial sUAS data collection and analytics. The goal of this book is to bridge the gap between complex technological concepts and practical applications, providing you with a general understanding of the fundamental principles, technologies, and workflows involved in drone technical services. This book is NOT a book about drone photography. If your desire is to learn artistic drone photography, this book will likely be of little benefit to you.

The rapid advancement of drone technology has revolutionized many industries, offering new capabilities for detailed and efficient geographical data collection. Drones, equipped with various sensors, provide unprecedented accessibility to aerial photogrammetry, surveying, and mapping techniques, which were once prohibitively costly and time-consuming.

This book is structured in such a way as to introduce you to the essential equipment and software used in drone mapping, and guide you through the processes that make effective drone mapping possible. From understanding the basic definitions and scopes of drone photogrammetry, to exploring the advanced techniques in data processing, this book covers a wide range of topics crucial for anyone looking to delve into the world of drone data collection.

It's important to acknowledge that while this book strives to provide the most current and comprehensive information on drone technologies, this field progresses so rapidly that some specifics regarding drones, sensors, and other tools mentioned may become obsolete before the book itself does. The industry's continuous innovation means that hardware and software evolve constantly. While specific equipment models may become outdated, the fundamental workflows and missions involved in drone mapping remain constant or change at a much slower pace. People will always need accurate maps, detailed surveys, and 3D models, and drones currently represent the fastest, most accurate, and most cost-effective method for obtaining this information.

This book is incomplete. Many times, in the text of this writing, you will see some variation of this comment. It is essential to note from the outset that there are too many drones, software packages, processes, and ancillary products to be able to discuss them all. Instead, the goal of this book is to give the reader a general overview and practical knowledge of how drones can be and are being employed in various aspects of the photogrammetry fields. When you have completed this text, you should have a solid understanding of how drones can benefit the world of photogrammetry, as well as a workable knowledge of how to employ this technology.

This book is redundant. You will read similar information over and over. Due to the nature of this field, it was decided to examine technologies in the context of how they are employed with the technology around them. This leads to multiple discussions about types of hardware, software, and workflows that may overlap, depending on how the technology is employed.

Whether you are a student, a budding professional in geospatial sciences, or someone with a keen interest in drone technology, this book should serve as a valuable resource. It provides practical insights and a solid foundation in drone mapping technology and techniques, ensuring you are prepared to handle projects that require high levels of accuracy and detail.

As the author, I wish to be very clear: while we discuss the potential legal ramifications of working in these fields, I am not a lawyer, a Professionally Licensed Surveyor, or a Certified Photogrammetrist, and I do not represent myself as being any of them in any way. I am a licensed drone pilot who has extensive experience in these fields, having worked with professionals in these fields since 2018. Now that that's clear, I must be very specific, do not take anything in this book as legal advice. ALWAYS check your local laws regarding all the subjects discussed in this book. Laws vary widely from place to place, so what may be legal in one place may be completely illegal in another. Don't assume you know, and do not take this book as legal advice; check with your own legal professional.

Thank you for choosing this book as your guide into the fascinating world of drone mapping. Let's embark on this journey together, exploring how cutting-edge drone technology continues to shape and redefine the capabilities of aerial photogrammetry.

Definitions and Scopes

This section is included to clarify key terms and concepts in drone mapping, surveying, and photogrammetry, providing a comprehensive understanding necessary for anyone new to the field. These definitions set the stage for more in-depth discussions in subsequent chapters and ensure that readers have a solid grasp of the fundamental terminology used in professional drone operations. Don't feel compelled to read all of these now, just know that this reference is here at the front of the book for the purpose of being able to quickly refer to needed information.

Absolute Accuracy: Absolute accuracy in drone mapping refers to the precision with which the mapped locations of objects or features correspond to their true positions on the Earth's surface, as defined by a global coordinate system such as GPS. This measure of accuracy is crucial for ensuring that the data captured by the drone is not only internally consistent (relative accuracy) but also correctly aligned with real-world coordinates. *See also: Relative Accuracy, Local Accuracy.*

AGL: (Above Ground Level) AGL refers to the altitude of an aircraft relative to the ground directly below it. For instance, if a drone were to take off at Dallas Love Field (KDAL) and hover at 10ft, it would be at

10ft AGL and 499ft MSL (Mean Sea Level) because KDALs elevation is 489ft. For the most part, manned aircraft only use AGL for cloud levels, virtually all other aspects of manned aviation rely on MSL.

Base Station: A fixed GNSS station that provides ground-based corrections for a drone's GPS signals. RTK (see definition) base stations like the Trimble, Emlid, (Absolute Accuracy) and DJI D-RTK series (Relative Accuracy) help drones achieve centimeter-level[1] (.4in) precision in navigation and data collection by correcting positional errors in real-time.

BVLOS, or Beyond Visual Line of Sight, refers to drone operations where the drone is flown outside the direct line of sight of the pilot or observer. Unlike visual line of sight (VLOS) operations, where the drone operator must keep the unmanned aircraft within their visual field, BVLOS permits the drone to travel further distances and be operated via remote controls and sensors. BVLOS operations require stringent adherence to regulatory guidelines to ensure safety and coordination within shared airspace. In the United States, a special permit known as a waiver is required for BVLOS.

Cadastral Map: A detailed map displaying the boundaries, ownership, and subdivisions of land parcels within a defined area, used primarily for legal, administrative, or taxation purposes. In drone mapping, these maps are generated or updated using aerial data collected by drones equipped with high-resolution cameras or LiDAR, capturing imagery or point clouds that are processed into precise geospatial representations of property lines, dimensions, and occasionally additional features like land use or structures.

CMOS Sensor (Complementary Metal-Oxide-Semiconductor): In digital cameras, CMOS sensors are semiconductor devices that convert light into electrons, functioning as the electronic eye of the camera. These sensors are built using CMOS technology, which allows for the integration of all necessary image sensor functions on a single chip, enabling faster processing, lower power consumption, and more compact designs compared to CCD sensors.

CORS Network: A Continuously Operating Reference Station (CORS) network is a system of permanently installed Global Navigation Satellite System (GNSS) ground-based receivers or ground stations that provide precise positioning data for a wide range of applications. These stations collect satellite signals and transmit the data to a central processing center, which is used to correct and enhance the accuracy of GPS positioning for users within the network's coverage area. CORS networks are crucial for high-precision applications, such as geodetic surveys, construction, agriculture, and autonomous vehicle navigation (such as used in farming). By offering real-time corrections and post-processing capabilities, CORS networks enable centimeter-level accuracy[1] and improve the reliability of positioning systems. Many states in the United States provide state-run networks to individuals and businesses at little or no cost. Most of these are run by state highway departments. In drone terms, these can function as part of an RTK network.

Drone: Referred to by the FAA as sUAS. The FAA (Federal Aviation Administration) defines sUAS (small Unmanned Aircraft System) as an unmanned aircraft weighing less than 55 pounds (25 kg) on takeoff,

---

[1]Although achieving 1cm absolute accuracy is feasible in highly controlled settings, it's not practical in real-world scenarios. Typically, the best achievable accuracy in most cases ranges from 3 to 5cm. For the majority of applications, an accuracy of 5cm to 10cm is generally considered sufficient

including everything that is on board or otherwise attached to the aircraft. This category includes the unmanned aircraft itself and all of the associated support equipment, communication links, and components that control the aircraft.

Federal Aviation Administration (FAA): The Federal Aviation Administration (FAA) is a governmental body of the United States responsible for regulating all aspects of civil aviation in the U.S. national airspace. Established in 1958, the FAA's primary roles include regulating air travel, certifying personnel and aircraft, setting standards for airports, and overseeing all components related to the safety of all civil aviation operations, including sUASs or Drones.

Focal length, actual: The actual focal length is the physical measurement of the distance between the lens and the image sensor when the lens is focused at infinity. It is a fixed and inherent property of the lens itself and doesn't change regardless of the camera or sensor. This measurement determines how much the lens can magnify the view of a scene and impacts the field of view. A shorter focal length provides a wider field of view, while a longer focal length offers a narrower view, effectively "zooming in" on distant subjects.

Focal length, equivalent: Equivalent focal length is a calculation used to compare the angle of view of a lens on a digital camera with a smaller sensor size to the angle of view of a lens with the same specified focal length on a full-frame or 35mm film camera. This is useful because many users are familiar with the perspectives and visual outcomes of lenses on standard 35mm cameras.

Geo-tag: A geo-tag is metadata embedded within an image or video file that usually includes the geographical coordinates (latitude and longitude) where the file was captured. It is often enriched with other data, such as altitude and the exact time of capture. In drone mapping, geo-tags are crucial for integrating aerial images with Geographic Information Systems (GIS) to create accurate maps and 3D models.

GIS (Geographic Information System): GIS is a framework used to gather, manage, analyze, and visually represent geographical and spatial data. In the context of drone mapping, GIS plays a crucial role in processing and utilizing the data collected by drones to create detailed and accurate geographic maps and 3D models.

GNSS (Global Navigation Satellite System): GNSS is a standard term for satellite navigation systems that provide autonomous geospatial positioning with global coverage. This system allows small electronic receivers to determine their location (longitude, latitude, and altitude) within a few meters using time signals transmitted along a line of sight by radio from satellites. Receivers that can use data from multiple systems offer improved accuracy and reliability simultaneously. Several countries around the world deploy their own GNSS system, for instance, the U.S. system is commonly referred to as GPS.

GPS (Global Positioning System): GPS is a global navigation satellite system maintained by the United States government and is freely accessible by anyone with a GPS receiver. The system provides geolocation and time information to a GPS receiver anywhere on Earth, with an unobstructed line of sight to four or more GPS satellites. This system relies on a constellation of satellites that transmit precise microwave signals, enabling GPS receivers to determine their location, speed, direction, and time. In addition to the American GPS System, several other countries operate similar constellations. They include

GLONASS: Russian Federation, Galileo: European Union, BeiDou: China, NavIC: India, and QZSS: Japan. *See Also: GNSS*

Ground Control Points (GPCs): GPCs are specific points on the ground with known coordinates used as reference points in aerial mapping. They are physically placed on the ground in such a way as to be easily visible in images captured by the drone. Ground control points are crucial for ensuring the accuracy and reliability of photogrammetric and Survey data when used with PPK technologies to adjust and calibrate the data collected by drones.

Ground Control Station (GCS): GCSs are the stations used to control a drone or group of drones. While this can refer to a Fixed Base of Operation, a Mobile Command Center, or a handheld remote control, in the context of this writing, it will most commonly refer to a handheld remote control.

Ground Sampling Distance (GSD): Ground Sampling Distance (GSD) refers to the distance between two consecutive pixel centers measured on the ground. It represents the actual ground distance covered by a single pixel in a drone-captured image. 1GSD = 1sqcm per pixel

LiDAR: Light Detection and Ranging is a remote sensing method that uses light in the form of a pulsed laser to measure variable distances to the Earth. In drone applications, LiDAR sensors mounted on UAVs scan the ground and collect height data, which is used to create precise, three-dimensional information about the shape and surface characteristics of the Earth. While it is generally believed that LiDAR is more accurate than Photogrammetry, in practice that isn't always the case. LiDAR is an excellent tool in areas where vegetation may be an issue, however, in open areas, many times, photogrammetry may be as good or an even better solution.

Local Accuracy: In mapping and surveying, local accuracy refers to the precision of a measurement within a specific area or localized region relative to other fixed measurements within the same location. It assesses how closely points or monuments within a localized area are positioned relative to one another without necessarily considering their positions in the broader geographic coordinate system. High local accuracy ensures that the relative positions of features or points within a surveyed area are consistently accurate. This is crucial for detailed mapping, construction, and engineering projects where the spatial relationships within a site are of primary importance.

Mapping: The act of creating maps, which are graphical representations of geographic areas, showing physical features and sometimes demographic or socio-economic data. Drone mapping refers to using drones to collect aerial data that is then processed to produce maps of areas, highlighting details not visible from the ground.

Monument: In surveying and civil engineering, a monument is a physical structure or marker that indicates a specific location on the ground. Monuments are used to define property boundaries, survey points, and other critical reference points. They can be natural objects like trees or rocks or man-made objects such as metal rods, concrete posts, or brass discs embedded in concrete. Monuments are essential for accurately and reliably identifying surveyed points over long periods, ensuring that the locations remain consistent despite changes in the surrounding environment.

MSL: (Mean Sea Level): is the average height of the ocean's surface, measured over a long period, which serves as a standard reference point for measuring elevation and altitude. This average is calculated by considering the fluctuations caused by tides, atmospheric pressure, and other factors over time, effectively smoothing out the short-term variations to establish a consistent baseline. MSL is used in various fields, including aviation, cartography, and meteorology, to determine heights and depths relative to this baseline. It is important to note that MSL can vary slightly depending on geographic location due to gravitational differences and local sea conditions.

Nadir (NAY-deer): The term "nadir" denotes the angle or direction of observation from a drone down to the Earth's surface, which is directly beneath the aerial platform, generally at a 90-degree down angle from the horizon. In aerial photography and drone surveys, capturing images from the nadir perspective involves positioning the camera to face directly downward. This perspective is crucial for tasks requiring precision and minimal perspective distortion, such as in topographic mapping, agriculture surveys, and other forms of geographic data collection.

Networked Transport of RTCM via Internet Protocol (NTRIP): An NTRIP network is a protocol used to stream differential GPS (DGPS) and real-time kinematic (RTK) correction data over the Internet. It facilitates the transmission of GNSS correction data from a base station or CORS network to a rover or user device, such as an RTK-enabled drone, via the Internet. This setup allows for high-precision positioning by correcting GPS signal inaccuracies in real time. NTRIP consists of a caster (server), client (rover), and source (base station) to manage the data flow.

NOTAM (Notice to Airmen or Notice to Air Missions) is a critical advisory issued by aviation authorities to inform pilots and air traffic controllers of any temporary changes or hazards in the National Airspace System (NAS) that could affect flight operations. This includes information on everything from closed runways, inoperable navigational aids, to special events like air shows or military exercises that might restrict airspace. NOTAMs are essential for ensuring safety in the skies by providing real-time updates that aren't known far enough in advance to be included in standard aeronautical publications.

Oblique: Pertaining to drone flight missions, oblique refers to imagery captured at an angle from the drone to the subject rather than directly overhead (nadir). This method is commonly used to capture the sides of buildings, slopes of terrain, and other vertical structures, offering a more comprehensive view that includes both the tops and the sides of objects within the same frame.

PPM: PPM, as referred to when reading the "advertised accuracy" of RTK on a drone, stands for "parts per million." It is a unit of measure used to describe the additional positional error that increases with distance. Specifically, 1 PPM means that for every one million units of distance (e.g., millimeters or meters), there is an additional unit of error. For instance, an accuracy specification of "1.5 cm + 1 ppm" indicates a base error of 1.5 centimeters plus an extra error that grows by 1 millimeter per kilometer of distance from the reference point or base station.

(Smart) Oblique: This is similar to an oblique mission; however, instead of the drone's gimbal being set at one angle during the flight and the drone flying some sort of cross-hatch pattern, the drone will fly a standard down-and-back pattern, and the camera will rotate both backward and forwards and side to side, capturing images. This allows for capture at multiple angles and a shorter, more efficient flight.

TFR (Temporary Flight Restriction): The Federal Aviation Administration (FAA) issues Temporary Flight Restrictions (TFRs) to control airspace temporarily for reasons like security, public safety, disaster response, VIP movements, or special aircraft operations. These restrictions are critical for ensuring safety and security, and pilots must check Notices to Airmen (NOTAMs) for current TFRs. Violating these restrictions can lead to severe penalties.

Orthomosaic Imaging: A detailed, accurate photo representation of an area created by stitching together a series of overlapping photographs taken from a drone. The resulting image maintains the true properties of the photographed area, providing a uniform scale and accurate data that can be used for measurement and analysis.

Payload: In the context of drones, the term "payload" refers to the equipment or cargo that a drone carries apart from its essential components required for flight, such as batteries, navigation systems, and control mechanisms. The payload typically includes cameras, sensors, communication devices, and sometimes even small items for delivery. The nature of the payload varies based on the drone's intended use, whether it's for photography, surveying, monitoring, or transport.

Photogrammetry: A technique of obtaining information about physical objects and the environment through the process of recording, measuring, and interpreting photographic images. Drone photogrammetry involves using drones to capture a series of overhead photos that are then used to create maps and 3D models.

Post-Processing Kinematic (PPK): Similar to RTK, PPK improves GPS data accuracy by collecting measurement data during the flight and then processing it afterward with corrections from a base station. This method is useful when real-time data transmission is difficult or higher processing flexibility is needed.

Radiometric Thermal: Radiometric thermal imaging is a technology that measures and records the temperature of objects, surfaces, or areas by detecting the infrared radiation they emit. Unlike standard thermal imaging, which only provides relative temperature differences as visual heat maps, radiometric thermal imaging captures absolute temperature data. This allows it to produce images where each pixel contains temperature measurements, which can be quantitatively analyzed.

Real-Time Kinematic: (RTK) is a satellite navigation technique that enhances the precision of position data derived from satellite-based positioning systems. This technique uses a single base station that provides real-time corrections, providing a much higher level of accuracy in the positioning of the drone as it collects data.

Relative accuracy: Relative accuracy in drone mapping refers to the precision of measurements when comparing one point to another within the same dataset. It does not take into account the absolute accuracy or the exactness of the location relative to real-world coordinates; instead, it focuses on how accurately the points are measured in relation to each other within the map or model created by the drone.

RGB: Short for Red, Green, and Blue. It is a color model in which these three colors are combined in various ways to produce a broad spectrum of colors in cameras. In the context of drone cameras, RGB is the standard format for capturing color aerial imagery. *The exception is Autel; the Autel Evo II V1 and V2 series of drones use an alternative technology called RYYB, which is Red, Yellow, Yellow, and Blue. There is a school of thought that RYYB sensors give a more natural feel to pictures taken in sunlight. Its known drawback, and why it is not widely adopted, is that RYYB cameras tend to be more sensitive to light and may wash out the image more quickly. While a few camera manufacturers also use this tech, Autel is the* *only drone manufacturer known to use it as of the publishing of this book. With the release of the Autel II V3 series of drones, the 6K one inch CMOS camera has been changed to a Sony RGB design. However, the Autel Evo Lite and Nano consumer drone series maintain the RYYB sensor type.*

RINEX File (Receiver Independent Exchange Format): A RINEX file is a standard format for raw satellite navigation system data. The format is used to store and exchange data collected from Global Navigation Satellite Systems (GNSS) such as GPS, GLONASS, Galileo, and BeiDou. These files are often used in post-processing GNSS data to achieve higher accuracy than real-time processing allows. The standardized format allows for the easy exchange of GNSS data between different receivers and processing software. They are essential in geodesy, surveying, and various other geospatial applications requiring precise positioning data.

Rover: In surveying, a rover is a portable GNSS receiver that is moved across the survey area to collect precise location data. It communicates with a fixed base station that continuously transmits correction data, allowing the rover to adjust and refine its calculated positions in real time or through post-processing. This setup is essential for establishing highly accurate ground control points (GCPs) in drone mapping and other geospatial applications.

sUAS (small Unmanned Aerial Systems): The FAA (Federal Aviation Administration) defines a sUAS (small Unmanned Aircraft System) as an unmanned aircraft weighing less than 55 pounds (25 kg) on takeoff, including everything on board or otherwise attached to the aircraft. This category includes the unmanned aircraft itself and all of the associated support equipment, communication links, and components that control the aircraft.

Survey: The process of determining precise locations and measurements on the Earth's surface, utilizing tools and techniques for collecting and analyzing geographic and environmental information. In drone applications, surveying involves using UAVs equipped with sensors to gather data about land, structures, and features.

Thermal Sensor: Thermal sensors are particularly valuable in drone mapping for identifying variations in land or building temperatures, detecting water stress in agriculture, finding thermal leaks in buildings, or observing wildlife in their natural habitats. They enable precise, non-contact temperature measurements over large areas or hard-to-reach locations, enhancing the efficiency and effectiveness of geological, environmental, and industrial surveys.

Volumetrics: volumetrics, often called cut/fill analysis, refers to the use of drones equipped with advanced sensors and imaging technologies to measure and calculate the volume of large objects or areas, such as stockpiles, pits, and landfills. This application is particularly prevalent in mining, construction, and agriculture, where accurate volume measurements are crucial for inventory management, planning, and regulatory compliance.

These definitions provide essential background knowledge for understanding the technologies and methods used in drone mapping; they are crucial for future discussions in later chapters.

Chapter 1: Introduction to Drone Mapping

The Basics:

Before diving into the technical aspects of drone mapping, it's crucial to address the legal framework that governs drone use, particularly when employed for purposes beyond recreational use. Understanding and adhering to these laws is fundamental to conducting drone operations responsibly and legally.

As drone technology continues to evolve and become more accessible, the opportunities for using drones in photogrammetry, mapping, and surveying are expanding rapidly. These tasks offer incredible potential for precision, efficiency, and cost savings across various industries, from construction and land development to environmental monitoring and agriculture. However, with these opportunities come responsibilities, and navigating the legal landscape is one of the most critical aspects of operating a drone professionally. Whether you're mapping a construction site, conducting a land survey, or capturing data for a 3D model, understanding the legal ramifications is essential for every drone pilot.

This section provides a basic overview of the legal considerations and regulations that you need to be aware of when using drones for photogrammetry, mapping and surveying. While this is not legal advice, and I am not a lawyer, the goal here is to help you identify the key areas where legal requirements and best practices intersect with your operations. From obtaining the necessary certifications and permissions to adhering to airspace restrictions and privacy laws, a solid grasp of these topics will not only help you stay compliant but also protect your business from potential legal pitfalls. As we examine these concerns, remember that the legal landscape varies widely by region and is always subject to change, so staying informed and consulting with legal professionals in your area of operations is a critical part of your journey into drone photogrammetry, mapping, and surveying.

Legal Requirements in the United States:

In the United States, operating a drone for any purpose other than the recreational exemptions outlined in USC 44809 requires adherence to specific regulations set forth by the Federal Aviation Administration (FAA) in Part 107 of the Federal Air Regulations (FAR Part 107). This means that any mapping, survey, photogrammetry, inspection, YouTube videos, wedding videos, high school football videos, or whatever you are doing with a drone likely requires you to begin by having a part 107 drone pilot license. The commonly used litmus test is "Are you making a profit?" That is not, however, how the law works. That thought process can quickly get pilots into trouble. Every drone pilot is required to have a Part 107 drone pilots license, unless they are operating under the exemptions laid out in USC 44809. These regulations are designed to ensure that drone operations do not interfere with manned aircraft, compromise public safety, or invade privacy. If you are utilizing a drone for photogrammetry, mapping, surveying, inspection, or any enterprise that is not strictly for your personal enjoyment, and you are doing so without a part 107 drone pilot's license, you are very likely violating Federal Air Regulations. If caught, you could face, at the minimum, civil penalties. Depending on the circumstances, those penalties could escalate to criminal charges, including felony criminal penalties. Even with a part 107 certificate, pilots are still required to verify and comply with all federal air regulations, one of the most important of which is airspace regulations. You must know these rules, including obtaining necessary waivers or clearances, registering your drone, and following operational guidelines that dictate when, where, and how you can fly.

International Legal Considerations: If you are reading this book from outside the United States, it is imperative to familiarize yourself with your local drone laws in your region. Drone regulations can vary significantly from one country to another, and what is permissible in one area might be restricted in another. Ensure that you consult local aviation authorities or legal experts to understand the specific requirements applicable to your location.

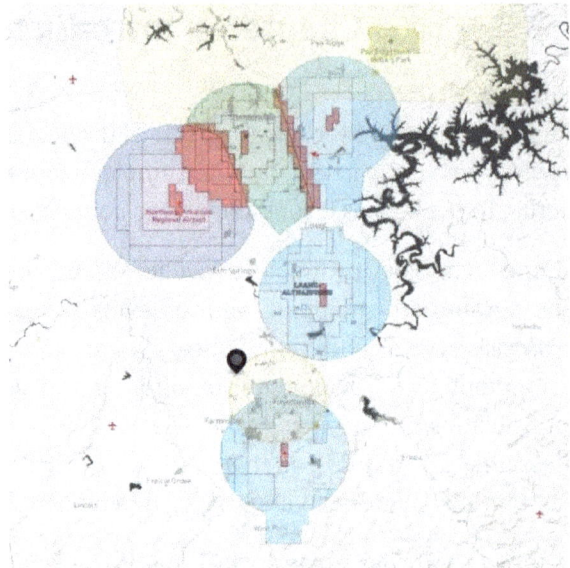

Legal Considerations in Drone Surveying and Mapping

Drone-based surveying and mapping unlock powerful geospatial capabilities, but their execution is governed by a complex legal landscape that varies across states and nations. This section outlines the regulatory framework, emphasizing the critical role of compliance and licensed surveyors in ensuring lawful, credible outcomes, particularly when data informs land disputes, construction, or official records.

Regulatory Landscape

Laws regulating surveying and mapping differ widely by jurisdiction, yet a common thread persists: specific qualifications or licenses are typically required for legal practice. In the United States, every state mandates licensure for surveyors, enforced through education, experience, and examinations, with similar standards in most countries. Mapping, while often less restricted, may still fall under professional oversight when tied to regulated outputs like cadastral maps. Non-compliance, whether through unlicensed surveying or misuse of mapping data, can trigger fines, legal disputes, or project delays, undermining both the operator and the data's integrity. Understanding these laws is as essential as mastering drone technology itself, safeguarding operations and their stakeholders.

The Role of Licensed Surveyors

Surveying carries heightened legal stakes, especially for property boundaries and construction, where precision and accountability are paramount. Licensed surveyors or civil engineers bring indispensable expertise to drone projects:

- **Compliance and Precision**: Their training, spanning 4-to-6-year degrees, years of fieldwork, and licensure, ensures adherence to regulations and delivers survey-grade accuracy, validated by tools like GPS, LiDAR, and drones.

- **Accountability**: Backed by professional liability insurance and ethical standards, they produce defensible results, answerable to licensing boards, unlike unlicensed operators whose errors risk costly repercussions.
- **Coordination and Risk Management**: They collaborate seamlessly with architects, engineers, and planners, mitigating legal and physical risks to protect projects and clients.
- **Technological Mastery**: Continuous education keeps them adept with evolving tools, ensuring data interpretation meets both technical and legal benchmarks.

Unlicensed drone-based surveying for official purposes—such as boundary delineation—is illegal, inviting penalties and eroding trust in the deliverables.

Implications for Drone Operators

For drone pilots, navigating this landscape demands more than technical skill—it requires legal awareness. Mapping missions, while versatile, must respect jurisdictional limits, particularly when outputs serve regulated ends. Surveying, however, demands licensed oversight; attempting it without credentials risks not just fines but professional credibility. A proactive approach—consulting regulations and partnering with surveyors—ensures operations are both effective and lawful, balancing innovation with responsibility. This diligence protects operators, clients, and the public, reinforcing drone technology's role as a credible geospatial tool.

Photogrammetry, mapping, and surveying are interconnected yet distinct disciplines critical to drone-based geospatial applications. Photogrammetry provides the image-based foundation, mapping creates visual representations, and surveying ensures precise measurements for legal and practical purposes. The legal complexities of surveying underscore the need for licensed professionals in drone projects, ensuring compliance, accuracy, and accountability. As drone technology advances, the synergy of these fields will drive innovation, enhancing efficiency and effectiveness in data collection and analysis.

Photogrammetry, Mapping, and Surveying with Drones

Drone-based geospatial applications rest on three interrelated disciplines—photogrammetry, mapping, and surveying—each harnessing aerial data to meet distinct yet complementary goals. This section defines their scope, details a general mission workflow from start to finish, and contrasts their execution, guiding operators in tailoring drone technology to project demands.

Core Concepts

- **Photogrammetry**: Extracts 3D spatial data from photographic images, triangulating positions from multiple angles. Rooted in Greek terms—*photos* (light), *gramma* (drawing), *metron* (measure)—it underpins maps and models, integrating with tools like LiDAR for versatility across aerial and terrestrial contexts.
- **Mapping**: Crafts visual representations of geographic areas, such as topographic or cadastral maps, using photogrammetry, GIS, and satellite data for navigation, planning, and resource management.
- **Surveying**: Measures precise 3D positions, distances, and angles on the Earth's surface, often for legal purposes like boundary delineation or construction, employing GPS, total stations, and drones.

These fields converge in their use of drone imagery but diverge in focus: photogrammetry generates data, mapping visualizes it, and surveying certifies precision for actionable outcomes.

General Mission Workflow

A drone mission unfolds through a structured lifecycle, adaptable to photogrammetry, mapping, or surveying objectives. Below, each phase is detailed, with recommendations for altitude, overlap, and other parameters serving as starting points—specifics can shift significantly based on project needs, site conditions, and desired outputs (explored further in Chapter 10).

Mission Planning

Every mission begins with defining objectives—e.g., a 3D model for construction, a map for land use, or a boundary survey—shaping the approach. Operators delineate the site (urban plot or rural expanse) using tools like satellite imagery or KML files in software such as DJI Pilot 2, setting waypoints for coverage. Recommended altitudes might range from 150-300 feet, targeting a Ground Sample Distance (GSD) of 0.5-1 cm/pixel, while overlap—often 70-80% front and side—ensures robust stitching. Capture modes (nadir, oblique, or both) align with goals, and flight timing accounts for weather and lighting, laying a precise foundation for data collection.

Flight Execution

With the plan set, drones launch on pre-programmed paths, capturing overlapping images tailored to the mission. A fleet like the Matrice 350 RTK might fly at 200 feet, snapping nadir shots every 0.7 seconds for orthomosaics or oblique shots at 45 degrees for 3D detail, adjusting for terrain or obstacles. Real-time monitoring via flight software tracks altitude, battery, and georeferencing signals (RTK or PPK), ensuring data integrity. Pilots may tweak speed or height mid-flight—e.g., lowering to 100 feet over complex structures—adapting to site-specific challenges while maintaining coverage and quality.

Data Processing

Post-flight, hundreds to thousands of images are ingested into software like Pix4Dmapper or Trimble Business Center. Processing identifies common features across frames, triangulating points into a sparse point cloud, refined by RTK (real-time corrections, 1-3 cm accuracy) or PPK (post-processed GNSS logs, sub-centimeter precision). Dense point clouds (millions of points), 3D meshes, orthomosaics, or digital terrain models (DTMs) emerge, with accuracy honed to benchmarks like <2-5 cm RMSE. This phase transforms raw captures into structured datasets, customized to project specifications through iterative adjustments.

Deliverables

The final stage crafts outputs for client use—orthomosaics as GeoTIFFs, 3D models as OBJs, or DTMs as LAS files—verified against georeferenced standards. A detailed report outlines methodology, equipment (e.g., drone model, sensor type), and metrics, ensuring clarity for applications like topographic analysis or legal filings. Delivery adapts to needs, from cloud links for planners to validated surveys for regulators, bridging technical precision to practical impact across industries.

Contrast and Comparison of Mission Types

While sharing this workflow, photogrammetry, mapping, and surveying diverge in execution and intent:

- **Photogrammetry Missions**:
    - *Purpose*: Extract versatile 3D data for models, maps, or analysis.
    - *Execution*: Balances nadir and oblique shots (e.g., 200 ft, 75% overlap) for comprehensive modeling, using RTK/PPK flexibly (1-3 cm accuracy). Outputs span meshes and orthomosaics, serving diverse needs without legal constraints.
- **Mapping Missions**:
    - *Purpose*: Visualize geographic areas for interpretation and planning.
    - *Execution*: Favors nadir shots (e.g., 300 ft, 70% overlap) for scalable 2D maps, with RTK/PPK ensuring alignment (<5 cm RMSE). Focuses on clarity over depth, requiring no licensure but technical skill.
- **Surveying Missions**:

- *Purpose*: Deliver precise, legally binding measurements for boundaries or construction.
- *Execution*: Blends nadir/oblique shots (e.g., 150 ft, 80% overlap) for exhaustive detail, demanding sub-centimeter RTK/PPK accuracy (1 cm + 1 ppm). Outputs like DTMs are validated by licensed surveyors for regulatory use.

Photogrammetry fuels data creation, mapping enhances visualization, and surveying ensures precision—each adapting the workflow to its unique stakes and outcomes, amplifying drones' geospatial potential.

Use Cases

In this section we're going to further explore some of the use cases where drones are having a significant impact on the cost and reliability of data gathering.

Environmental Monitoring

Environmental monitoring is critical for understanding and managing natural resources, ecosystems, and ecological changes. Drones offer an effective solution for monitoring large areas quickly and efficiently.

- **Habitat Mapping and Wildlife Monitoring**: Drones equipped with high-resolution cameras and sensors can capture detailed images of habitats, allowing for accurate mapping and monitoring of wildlife populations. This helps track changes in animal habitats, migration patterns, and population dynamics.

- **Climate Change Research**: Drones are used to monitor glaciers, ice caps, and other sensitive environments impacted by climate change. They provide data on melting rates, ice thickness, and changes in land cover, contributing valuable information to climate research.

- **Coastal and Wetland Monitoring**: Drones can map and monitor coastal and wetland areas, providing data on erosion, sediment transport, and vegetation health. This information is crucial for managing these vulnerable ecosystems and planning conservation efforts.

Inspection of Large Infrastructure

Large infrastructure projects, such as bridges, dams, and highways, require regular inspections to ensure structural integrity and safety. Drones offer a safer and more efficient method for conducting these inspections.

- **Bridge Inspections**: Traditional bridge inspections often require lane closures and scaffolding, which poses risks to workers and causes traffic disruptions. Drones can inspect hard-to-reach areas of bridges, capturing high-resolution images and videos to identify potential issues such as cracks, corrosion, or structural weaknesses.

- **Dam Inspections**: Dams are critical infrastructure that requires regular monitoring. Drones can survey the entire structure, including the upstream and downstream faces, spillways, and surrounding areas. This allows for early detection of potential problems like seepage, erosion, and structural damage.

- **Highway and Roadway Inspections**: Drones can quickly survey long stretches of highways and roadways, identifying potholes, cracks, and other surface damage. This enables maintenance teams to prioritize repairs and improve road safety.

Power Infrastructure

Power infrastructure, including power lines, solar farms, and windmills, is essential for modern society. Drones provide a cost-effective and efficient method for inspecting and maintaining these critical assets.

- **Power Line Inspections**: Inspecting power lines traditionally involves helicopters or climbing, both of which are risky and expensive. Drones with high-resolution cameras and thermal sensors can inspect power lines for damage, corrosion, and vegetation encroachment, ensuring reliable power delivery and reducing outage risks.

- **Solar Farm Inspections**: Solar panels need regular inspections to ensure they function efficiently. Drones can quickly survey large solar farms, using thermal cameras to detect faulty panels, hotspots, and shading issues. This helps maintain optimal performance and maximize energy production.

- **Windmill Inspections**: Wind turbines are often located in remote and hard-to-reach areas. Drones can inspect wind turbines, capturing detailed images of the blades, nacelle, and tower. This helps identify issues such as blade cracks, lightning strikes, and wear and tear, ensuring the turbines operate efficiently.

Forestry Management

Forestry management involves monitoring forest health, managing resources, and assessing damage from natural disasters. Drones offer an efficient way to gather data over large forested areas.

- **Forest Health Monitoring**: Drones equipped with multispectral and thermal cameras can assess the health of trees by detecting stress, disease, and pest infestations. This information is crucial for forest managers to take timely action and maintain forest health.

- **Timber Inventory and Management**: Drones can create detailed maps of forest areas, helping estimate timber volumes and plan harvests. This improves the efficiency of timber management and ensures sustainable practices.

- **Fire Management and Damage Assessment**: Drones can be deployed to monitor wildfires, providing real-time data on fire spread and intensity. After a fire, drones can assess the damage, helping in recovery efforts and planning for reforestation.

Crop Management

In agriculture, drones play a vital role in precision farming by providing detailed data on crop health, growth, and yield potential.

- **Crop Health Monitoring**: Drones equipped with multispectral cameras can capture images that help assess crop health. They can detect issues such as nutrient deficiencies, disease, and water stress, enabling farmers to take corrective actions promptly.

- **Field Mapping and Planning**: Drones can create high-resolution maps of fields, helping in planning planting patterns, irrigation systems, and soil management practices. This leads to more efficient use of resources and higher crop yields.

- **Pest and Weed Management**: By identifying areas affected by pests and weeds, drones help target treatments more accurately, reducing the use of pesticides and herbicides and minimizing environmental impact.

Other Applications

Drones have a wide range of other applications in various fields, showcasing their versatility and utility.

- **Disaster Response and Management**: Drones are used in disaster response to assess damage, locate survivors, and deliver supplies. They provide real-time data to aid in decision-making and coordinate rescue efforts.

- **Urban Planning and Development**: Drones provide detailed aerial imagery and 3D models essential for urban planning and development. They help assess land use, plan infrastructure, and monitor construction progress.

- **Archaeological Surveys**: Drones offer a non-invasive method for surveying archaeological sites. They can capture high-resolution images and 3D models, helping archaeologists map and document sites without disturbing the ground.

Drones have become indispensable tools in photogrammetry, mapping, and surveying,, offering numerous advantages in efficiency, safety, and accuracy. From environmental monitoring and infrastructure inspection to forestry and crop management, drones provide valuable data that enhances decision-making and resource management. As drone technology advances, their applications will undoubtedly expand, further cementing their role in various industries and fields.

Surveyors:

The Guys Who Hold the Sticks

While this section may seem somewhat redundant based on what has been previously discussed in this chapter, it is important to note that, while geospatial analytics from the perspective of a drone or aircraft are relatively new technologies, surveying has taken place for virtually all of recorded history. Surveyors are a unique breed to say the least. Because of that, it is prudent to take a deeper dive into the history and the Culture surrounding surveying, which, until a few hundred years ago was synonymous with civil engineering.

Surveying is one of the oldest professions in the world, with its practice dating back thousands of years. The need for surveying has been apparent throughout human history as civilizations required the demarcation of land for ownership, construction, and agriculture.

Historical Context

1. **Ancient Egypt (circa 2700 BC):** Surveying techniques were used during the construction of the Great Pyramids. Ancient Egyptians employed rudimentary surveying tools to plan, design, and ensure the precise alignment and orientation of the pyramids with respect to the cardinal points.

2. **Canaan (circa 1250 BC):** Joshua 18:6 NKJV: "You shall therefore survey the land in seven parts and bring *the survey* here to me, that I may cast lots for you here before the Lord our God." This is the first direct reference to survey in Biblical scripture. However, it is inferred that ancient Israel performed survey work even while in Egypt and may have learned the technology from the Egyptians.

3. **Babylon (circa 1200 BC):** The practice of surveying land parcels for agricultural and legal purposes was common in Babylon. Evidence from the Code of Hammurabi shows rules regarding boundary markers and land measurement.

4. **Ancient Greece (circa 500 BC):** Greeks advanced the science of surveying, incorporating geometric principles and mathematics more significantly into land measurement. Prominent Greek philosophers like Thales and Pythagoras contributed to these developments, and the concept of "Geometry" itself, which translates to "earth measuring," underscores its close ties with surveying.

5. **Roman Empire (circa 300 BC - 500 AD):** The Romans were known for their engineering and construction, relying heavily on surveying. They developed tools like the groma and dioptra to construct roads, aqueducts, and buildings, extending surveying techniques across their vast empire.

Medieval to Modern Transition

- **Middle Ages (500 AD - 1500 AD):** Surveying played an essential role in the feudal system for managing and dividing land, although many technical advances from the Romans were temporarily lost during the early medieval period.

- **Renaissance (14th - 17th Century):** The resurgence of learning during the Renaissance included the revival and advancement of surveying. This period saw significant developments in cartography and navigational technologies spurred by exploration and the Age of Discovery.

Modern Era

- **18th Century Onwards:** The profession became more formalized during the 18th Century with the establishment of more precise instruments and techniques. The invention of the theodolite, which allowed for more accurate angle measurements, was a significant advancement.

- **19th and 20th Century:** Industrialization and the expansion of colonial empires increased the need for detailed and accurate land surveys worldwide. During this time, land surveys were crucial for building railways, mapping national boundaries, and developing new territories.

- **21st Century:** Today, surveying incorporates advanced technologies such as GNSS (Global Navigation Satellite Systems), GIS (Geographic Information Systems), drones, and LiDAR (Light

Detection and Ranging) to achieve highly accurate land measurement and analysis. These tools allow surveyors to perform their tasks more efficiently and precisely than ever before.

From the construction of ancient monuments to modern-day urban planning and development, surveying has been an essential activity, continuously evolving with each era of human advancement. Its historical lineage reflects the continuous human effort to understand and organize the physical space for various administrative, legal, and construction purposes.

Surveying: Precision and Professional Responsibility

Surveying stands as a cornerstone of land development and infrastructure, a profession steeped in millennia of history yet vital to modern drone mapping. For those eager to fly drones and produce deliverables like boundary maps, topographies, or volumetric analyses, understanding surveying's role, and its legal boundaries, is non-negotiable. Without a surveyor's license, civil engineering credentials, or employment under such a professional, claiming to "survey" is illegal across every U.S. state and most countries. This section explores why surveyors remain indispensable, detailing their workflow and underscoring the expertise they bring to drone-based projects.

The Surveyor's Craft

Surveying demands a rigorous blend of education, training, and practical skill. Professionals typically hold a 4-to-6-year degree in surveying, geomatics, or civil engineering, followed by up to six years of on-the-job training in some regions. Licensure, required universally in the U.S. and many nations, involves passing exams and maintaining competency through ongoing education, ensuring surveyors meet stringent standards of accuracy and ethics.

Each project begins with a clear grasp of its objectives: delineating property lines, mapping terrain for construction, or assessing environmental features. In the field, surveyors deploy tools ranging from traditional theodolites to modern GPS, total stations, and drones, measuring angles, distances, and elevations with precision. Multiple readings verify data integrity, recorded meticulously via tablets or laptops for real-time analysis. Office work follows, where raw measurements are processed into detailed maps and plans using CAD software, accompanied by reports that document methodology and findings for stakeholders like architects or legal authorities.

Collaboration and Specialization

Surveyors bridge technical data to practical outcomes, consulting with engineers, planners, and construction teams to align measurements with project needs. They may also serve as expert witnesses in boundary disputes, their work carrying legal weight. Specializations like land surveying for boundaries, topographical for contours, hydrographic for underwater features, or engineering for construction, tailor their expertise to diverse applications, all enhanced by technologies like GNSS, LiDAR, and UAVs, which they adopt through continuous learning.

Relevance to Drone Operators

Surveying's precision and accountability set it apart, rooted in a legacy from ancient Egypt to today's GNSS-driven era. For drone pilots, this history isn't mere trivia—it's a reminder of the profession's gravity. Surveyors ensure developments are built correctly, legally, and safely; a responsibility no unlicensed operator can assume. Entering a site with the intent to "show them how it's done" risks rejection; instead, collaboration with surveyors leverages their authority and knowledge. Whether on day one or day one thousand, their expertise, honed over years and validated by law, offers lessons in precision and professionalism, a humbling counterpoint to overconfidence. Proverbs 16:18 – Pride comes before a fall, and a haughty spirit before stumbling.

Historical Evolution and Future of Drone Mapping

Drone mapping's ascent reflects a dynamic interplay of aviation, imaging, and computational advances, evolving from rudimentary aerial experiments to a cornerstone of modern geospatial analysis. This journey, brief compared to surveying's ancient roots, has reshaped how we interpret and manage our environment, delivering precision and accessibility across industries. Below, we trace its development and peer into its promising future, highlighting milestones and trends that underscore its growing indispensability.

From Balloons to Drones

The origins of aerial mapping date to the mid-19th century, when gas balloons lifted cameras to capture the first overhead views. By the early 20th century, kites, pigeons, and airplanes expanded this scope—World War reconnaissance honed aerial photography's strategic value, later adapting to civilian needs like topography and urban planning. Modern drone technology emerged in the late 20th century, born from military intelligence-gathering UAVs. By the early 2000s, civilian adoption surged, driven by affordable, GPS-enabled drones paired with lightweight, high-resolution cameras, marking the dawn of accessible aerial mapping.

Technological Milestones

The integration of photogrammetry and LiDAR catalyzed drone mapping's transformation. Photogrammetry—merging overlapping images into maps and 3D models—offered a cost-effective way to survey vast or rugged terrain, while LiDAR's adaptation to drones delivered precise topographic data through dense vegetation, revolutionizing forestry, geology, and planning. Software advances in the 2010s amplified this shift: tools like Pix4D and Agisoft Metashape automated image stitching and data processing, incorporating AI to enhance accuracy. Meanwhile, regulatory frameworks matured, with rules governing airspace and privacy shaping safe, professional operations—foreshadowing expansions like the anticipated Part 108 for beyond-visual-line-of-sight (BVLOS) flights.

Current Impact and Capabilities

Today, drone mapping excels in gathering data swiftly and accurately, supporting disaster response, environmental monitoring, construction, and more. Enhanced cameras with larger sensors capture finer details across varied conditions, while compact, efficient LiDAR broadens topographic precision. Software now processes thousands of images with machine learning, reducing errors and accelerating insights. These advancements, paired with drones' ability to carry multispectral or thermal sensors, tailor data to specific needs, cementing their role across diverse sectors.

Future Horizons

Drone mapping's future shines bright, propelled by innovation and integration. Anticipated regulations (e.g., Part 108) will enable BVLOS missions and larger UAVs, expanding scope for infrastructure and environmental projects. Autonomous drone swarms promise to survey vast areas efficiently, while advanced sensors—multispectral, hyperspectral—and AI will deepen real-time analysis, from crop health to predictive modeling. Collaboration with satellite and ground data will yield richer, multi-layered maps, enhancing decision-making in urban planning and climate studies. As ethical and regulatory frameworks evolve to address privacy and airspace, drones will remain vital, pushing geospatial boundaries with precision and scale.

This evolution—from balloons to autonomous systems—positions drone mapping as a dynamic partner to photogrammetry and surveying, its trajectory poised to redefine geospatial intelligence in the chapters ahead.

Chapter 2: Drones

An important note from the author:

As a caveat to this chapter, as with much of this book, this writer would be remiss if it were not mentioned that this work is incomplete. As part of creating this book, and for what should be obvious reasons, the decision was made to use examples from a few different manufacturers to demonstrate the differences that may exist in different types of hardware and software. There are hundreds of drones on the market, at least as many sensors, and potentially as many as one hundred data processing suites available. No compensation for any product named in this book has been received, nor has any manufacturer or developer requested to be in it, and only a few have even been made aware of the publishing of this book. The author is writing from personal experience and publicly available data when publishing information about this technology. It is up to the reader to research before purchasing a drone for an intended mission or seek professional consultation if they are unsure.

One more thing

As the author of what could effectively be construed as a textbook about drones, I strive not to "break the fourth wall." It is my desire to write in a clear and objective manner. Though this will, for the most part, be the case, I want to take one paragraph of this book and personally address an issue that is rampant in the drone community, especially on social media. That is, well-intentioned people giving terribly inaccurate and misleading advice! So, Herein, I will give some of my own; if you go to the internet, be it Facebook, Instagram, or YouTube, remember, there is much advice being given by those who lack information and experience, and even more by those who simply seek to sell you something. So, take everything you read with a grain of salt. As the author of this book, I strived to research every aspect of this text so that I know the information I provide the reader is as thorough and accurate as possible. If I make a mistake, it is not intentional, and I will be quick to admit it, address it, and correct it. That being the case, I would like to address one of the most significant pieces of bad advice out there, which is, "You can map with any drone." It is literally the modern equivalent of "just any ole drone will do." While that is technically correct, it is the modern equivalent of "just saddle the horses." Because, while technically, you can ride a horse from the Atlantic to the Pacific, in the age of modern travel, it's just not feasible. Buying the wrong drone can be costly and very detrimental to your future prospects in the drone industry. In the following chapters, we cover the reasons for and the technology behind what are generally considered good and bad choices. If you are ready to buy today, and mapping, survey, or photogrammetry are going to be a part of your business plan, please do not rush out and buy just any consumer drone. Here are the keys to look for:

Consumer vs. Enterprise Drones: Suitability and Technology for Specialized Aerial Mapping

In the rapidly evolving world of unmanned aerial vehicles (UAVs), distinguishing between consumer and enterprise drones is crucial for professionals involved in specialized applications such as mapping, photogrammetry, and survey. This chapter will explore the key differences between popular consumer drones, like those from DJI and Autel, and their enterprise counterparts. It also highlights the reasoning for selecting enterprise-level drones for professional applications, emphasizing the role of native mapping software, available Software Development Kits (SDKs), third-party software support, and vastly superior sensors that enhance drone capabilities. While some first and second generation prosumer drones blur the line between consumer and enterprise, the divide in every sector is widening. Drones like the Phantom

4 Pro series, The DJI Mavic 2 series and Air 2s, and the Autel Evo II have seen some success in both realms, in more modern drones such as the Mavic 3, Mavic 3 Classic, Mavic 3 Pro, the Mini 4 Pro, and the Air 3, the lack of mechanical shutters, and especially, the lack of RTK[2], native mapping software or SDKs[3], have rendered them all but useless for photogrammetry related fields.

Just for Reference:

When discussing the DJI Mavic 3/Anzu Raptor series of drones, it is essential to note that there are eight versions of this drone under the DJI brand alone. There are five consumer or "prosumer versions," including the Mavic 3, Mavic 3 Cine, Mavic 3 Classic, Mavic 3 Pro, and the Mavic 3 Pro Cine. These drones are defined by their use of the DJI Fly app for control, lack of a mechanical shutter, lack of an RTK connection port, and lack of native mapping software or third-party SDK support.

The enterprise versions, including the Mavic 3 Enterprise, Mavic 3 Enterprise Thermal, and the Mavic 3 Multispectral, all support a wide variety of mapping functions including RTK, mechanical shutters (in the case of the Mavic 3 Enterprise and Mavic 3 Multispectral), available SDKs, and the use of the DJI Pilot 2 App for flight operations.

The RC Pro and RC Pro Enterprise controllers pictured above appear identical but are functionally very different.

It is also important to note that while the Mavic 3 Pro and the Mavic 3 Pro Cine have an option to ship with the DJI RC pro controller, it is a different model than the RC Pro Enterprise controller used on the Enterprise versions of the Mavic 3. The RC Pro version used on the consumer versions of the Mavic 3 employs the OcuSync 3 transmission system and the DJI Fly app. This unit's model number is RM510. The Enterprise versions of the RC Pro use the OcuSync 3 Enterprise transmission system and the much more advanced DJI Pilot 2 app. The model number on the enterprise version is RM510B. Contrary to what you may read online, these RC Pro controllers are not readily interchangeable. While it is possible, if you have the skillset and some very special software, to flash the firmware of either one to operate in the capacity of the other, these controllers where not meant to be interchanged. Switching them will result in extremely limited functionally and no ability to upgrade.

Understanding Drone Categories

Consumer Drones

Consumer drones, such as the DJI Air 3 / 3s, Mini 4 Pro, Mavic 3 Pro, Autel Evo Lite Plus, and Autel Nano, are primarily designed for general use by hobbyists and photography enthusiasts. These drones are characterized by their:

---

[2] RTK or Real Time Kinematic, is a GPS technique used to produce hyper accurate data results. See Glossary
[3] An SDK is a "Software Development Kit – a software Package released with some drones allowing 3rd party developers to create applications that will function directly with the drones. This is most often used for third party mapping software.

Affordability: They are generally more cost-effective, making them accessible to a broad audience.

Ease of Use: These drones are designed with user-friendly interfaces and are suitable for beginners.

Compact Design: Many models, like the DJI Mini 4 Pro and the Autel Evo Nano Plus, are lightweight and portable, ideal for travel and recreational photography.

DJI Air 3

Limited Customization: Consumer drones offer limited hardware modifications and software integration options.

Enterprise Drones

Contrasting sharply with consumer models, enterprise drones like the DJI Mavic 3 Enterprise, Autel Evo II Enterprise V3, the Inspired Flight IF800 Tomcat, and the Vision Aerial Switchblade Elite are engineered for professional use, including:

Vision Aerial Switchblade Elite Tricopter

Advanced Capabilities: These drones have superior cameras, thermal imaging, multispectral, hyper-spectral, and LiDAR sensors available. Many have longer flight times and carry heavier payloads, which is essential for detailed data collection. Virtually all will include cameras with mechanical shutters (a game changer in photogrammetry) and native mapping software.

RTK Enabled: In drone engineering and georeference-related work, RTK-enabled drones are all but necessary. In the modern connected world, relative, absolute, and local accuracy work together to achieve the most precise results. Using a drone that lacks RTK puts the pilot at a severe disadvantage.

Mechanical or Global Shutter: A common misconception within the drone industry is that a global shutter and a mechanical shutter are basically the same. That cannot be further from the truth. While these terminologies are sometimes confused, such as on the DJI Zenmuse P1, they are different. Regardless of which drone you choose to purchase, it should be equipped with at least one of these two technologies if it will be used for mapping, survey, or photogrammetry. Later in this book, all types of sensors and shutters are covered in great detail, giving you the insight to determine what would best fit your needs.

Enhanced Durability: Built to withstand varied environmental conditions, making them reliable for frequent and intensive use.

Extended Customization and Integration: They support various add-ons and modifications in software and hardware that are crucial for specific tasks.

The DJI Matrice M350RTK (pictured above) is IP55 Rated, has several different payload options, including thermal, LiDAR, Photogrammetry, and gas detection, and includes one of the most comprehensive mapping packages available today.

Mapping Software: Unlike most consumer drones, which lack automated flight software or the SDKs to create it, virtually all enterprise drones include extensive mapping and flight automation software and are compatible with many third-party apps.

Regulatory Compliance: Enterprise drones are often designed to meet specific regulatory standards for commercial, industrial, or government operations.

Importance of Enterprise Drones in Specialized Applications

For applications such as mapping, photogrammetry, and survey, enterprise drones are not just beneficial but often necessary due to:

Higher Precision and Quality: Enterprise drones typically have better sensors and cameras that produce high-resolution images with better depth of field and more accurate GNSS, which are crucial for accurate map creation.

Robust Software Support: They are compatible with advanced mapping software solutions like DroneDeploy, Pix4D, and DroneLink, which are essential for collecting aerial data and generating precise maps.

Greater Payload Capacity: Many allow the attachment of specialized equipment necessary for various mapping activities, such as high-precision GNSS, high-resolution photogrammetry cameras, LiDAR sensors, and other specialized devices such as multispectral, hyperspectral, and thermal.

The Role of SDKs in Enhancing Drone Functionality

A Software Development Kit (SDK) is a set of software tools and programs developers use to create applications for specific platforms. In the context of drones, it is used for:

Rock Robotic R3Pro V2 LiDAR unit with both SLAM and Photogrammetry abilities

Customization: SDKs allow developers to build custom applications that can control drones beyond the standard software, enabling tasks specific to mapping surveying, or other detailed applications that are not supported out of the box.

Integration: With SDKs, drones can be integrated into broader ecosystems, allowing for synchronization with other digital tools and improving workflow efficiencies. For example, by using software such as fleet management systems by companies like Aloft and DJI, project managers can monitor drone progress in real-time, adjusting resources to complete projects faster and more accurately.

Importance of Third-Party Support

The adoption of third-party software like DroneDeploy, Pix4D, and DroneLink is critical in the mapping industry because:

Enhanced Data Processing: These applications provide powerful tools for processing and analyzing drone-captured data, turning raw images into detailed maps and actionable data.

Scalability: Third-party applications can often handle data from multiple drones or larger datasets more efficiently than default software, which is essential for large-scale projects.

Specialized Features: Enterprise drones often include features tailored to specific types of surveying, mapping, or inspection work, which might not be available in the default software suite. This may include specialized 3D Ortho mapping of windmills or cellular towers, solar farm inspections, and power line mapping.

Choosing the correct type of drone is paramount for professionals in mapping, photogrammetry, and survey. While consumer drones offer ease of use and affordability, enterprise drones provide the robustness, precision, and customization required for professional applications. The importance of SDKs and third-party software in extending the capabilities of these drones cannot be overstated, as they enable the integration of advanced mapping technologies that are essential for accurate, efficient, and scalable outcomes. Therefore, investing in enterprise drones equipped with robust SDK and third-party support is crucial for professionals leveraging aerial technology for sophisticated mapping tasks.

As Chris the Drone Geek would say
"Let's Talk Drones"

Drone mapping has revolutionized industries by enabling rapid, efficient, and accurate geospatial data collection. This section introduces the key drone systems used in photogrammetry, mapping, and surveying, laying the groundwork for later chapters on sensors, positioning systems, flight management, and software. By understanding these platforms, readers will gain a solid foundation to master drone-based workflows.

Types of Drones Commonly used in mapping and inspection operations.

Enterprise drones, designed for commercial applications beyond cinematography or photography—such as thermal inspection, multispectral analysis, search and rescue, agriculture, mapping, and surveying—come in two primary designs: fixed-wing and multirotor platforms. These are further categorized by payload configurations: modular systems with interchangeable payloads or drones with fixed payloads. Let's explore their characteristics, benefits, and use cases in mapping and related fields.

Fixed-Wing vs. Multirotor Drones: Characteristics and Applications
In the world of unmanned aerial vehicles (UAVs), fixed-wing and multirotor drones offer distinct designs and capabilities, each tailored to specific tasks in photogrammetry, mapping, and surveying.

Design and Mechanics:

Fixed-Wing Drones: Resembling traditional airplanes, fixed-wing drones feature rigid wings that generate lift through forward motion, powered by a propeller (electric or jet-driven). They cannot hover and require continuous forward

WingtraOne Gen II

Autel Dragonfish Pro

movement to stay aloft, often needing launch and recovery systems like catapults or runways—though some newer models use tilt rotors for vertical takeoff and landing (VTOL), primarily for launch and recovery, lacking the stability of multirotors in tight spaces. These drones are generally larger, faster, and costlier, making them ideal for large-scale missions covering thousands of acres.

Multirotor Drones: Including tricopters, quadcopters, hexacopters, and octocopters, multirotors lift off vertically using multiple propellers on a horizontal plane, enabling hover, precise maneuvers, and agile flight in all directions. Typically smaller and more affordable than fixed-wing drones, they excel in stability and versatility, particularly in complex environments.

Inspired Flight IF800 Tomcat

Flight Dynamics and Capabilities:

Fixed-Wing Drones: Boast long endurance (up to five hours or more), high speeds (over 50 mph or 80 km/h), and efficiency, ideal for covering vast areas like agricultural fields or large surveying sites. Their forward motion, however, makes them susceptible to crosswind yaw, limiting their use with complex sensors like LiDAR. They typically carry heavier payloads, suiting them for professional-grade photogrammetric mapping equipment, but their design restricts operations in confined or urban areas.

Multirotor Drones: Offer shorter flight times (typically 30–50 minutes) but excel in maneuverability, stability, and stationary flight, crucial for precise tasks like aerial photography, LiDAR mapping, and low-altitude photogrammetry. Less affected by crosswind yaw due to their horizontal rotor configuration, they're better suited for complex sensors. Their vertical takeoff and landing (VTOL) capability eliminates the need for runways, making them perfect for urban, wooded, or mountainous operations. Many multirotors, like the DJI Matrice 350 RTK, feature modular or multi-payload systems, allowing quick deployment with a variety of sensors (e.g., up to three payloads simultaneously).

Applications:

Fixed-Wing Drones: Their range and endurance make them ideal for large-scale aerial mapping and surveying, efficiently covering thousands of acres for topographic surveys, GIS mapping, and agricultural planning using photogrammetry. They're also used in environmental monitoring, carrying sophisticated sensors for wildlife tracking or ecosystem analysis. While not as well suited for LiDAR due to cross wind yawing tendencies, their long-range capabilities align with other technologies Such as long-Range Delivery.

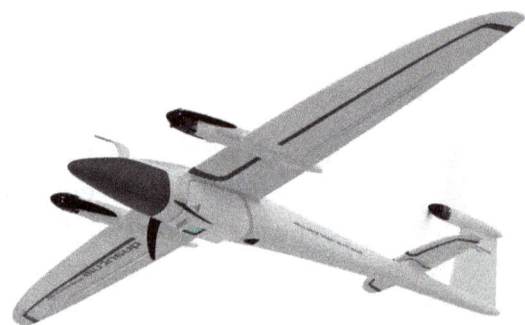
Quantum Systems Trinity F90+

      While not directly related to Mapping, Survey, and Photogrammetry, it is essential to note that companies like Zipline Inc. have worked diligently to make long-range delivery a reality. Operating in remote parts of Africa, delivering life-saving medications to remote facilities, they are now bringing that expertise to the United States intending to use US operations to fund expansion in the Third World. Operating under FAR part 135 in the United States and approved for BVLOS operations, Zipline drones routinely perform deliveries of up to 4.4 lbs at distances of up to 12.5 miles. Their next generation of drones, due to begin testing at their Pea Ridge Arkansas facility in Q3 2024, will have range capabilities over 50 miles, will require less ground crew, and will be capable of delivering packages on targets the size of a dinner plate. Though 50-mile ranges are not currently planned for operations, they are breaking new ground in delivery.

Multirotor Drones: Their agility and stability suit them for LiDAR operations, small to medium photogrammetry missions (up to ~500 acres, though capable of over 5,000 acres daily under optimal conditions), and detailed inspections of infrastructure like bridges, solar farms, power lines, and wind turbines. They're also ideal for videography and cinematography, leveraging hover and tight maneuvers for dynamic imaging in real estate, events, or film. Above 500 acres, fixed-wing drones often become more practical, but multirotors dominate smaller, complex, or urban projects.

Vision Aerial Switchblade Elite

Understanding these differences empowers drone operators to choose the right platform for their photogrammetry, mapping, or surveying needs, balancing cost, coverage, and precision for optimal results.

Advantages and Disadvantages

Fixed-Wing Drones

Advantages: Longer flight times, greater speed, and larger payload capacity.

Disadvantages: They require more space for takeoff and landing, are less maneuverable, and are generally unsuitable for tasks requiring stationary hovering. They are also more susceptible to crosswind and generally have lower wind tolerance.

Multi-Rotor Drones

Advantages: High maneuverability, VTOL capabilities, and excellent stability in all phases of flight.

Disadvantages: Shorter flight times due to higher energy use and typically limited by smaller payload capacities.

The choice between fixed-wing and multi-rotor drones largely depends on the specific requirements of the task. Fixed-wing drones are unmatched in efficiency and endurance for large-scale mapping and monitoring tasks. In contrast, multi-rotor drones offer unparalleled precision and flexibility for LiDAR, detailed inspections, and aerial imaging. Understanding these differences helps users select the suitable drone to meet their operational needs effectively, ensuring successful outcomes across various applications.

In discussing the differences between fixed-wing and multi-rotor drones, it's important to note that several newer designs on the market are, in essence, hybrid drones; they are a combination of a fixed-wing in flight and a multi-rotor during takeoff and landing. Most have a rotating wing that is vertical for takeoff and landing and transitions to horizontal for flight, or they are equipped with an additional set of horizontal rotors to transition from one phase of flight to another. Though they are an improvement over standard fixed-wing drones in the takeoff and landing phases of flight, they continue to suffer from the drawbacks of fixed-wing drones while in the mission phase of flight and add the burden of additional weight and more complexity. For this reason, they are generally regarded as fixed-wing drones in the context of this book.

Modular Systems (Changeable Payloads)

A modular or changeable payload on a drone refers to a configuration where the equipment carried by the drone—such as cameras, sensors, or other data collection instruments—can be easily swapped or replaced according to specific mission requirements. This feature allows for various types of payloads to be attached to the drone, or sometimes up to 3 at a time, depending on the needs of the particular task, such as thermal imaging cameras, multispectral sensors, LiDAR units, or high-resolution cameras.

Drones with modular or interchangeable payloads offer significant flexibility, a major advantage in various commercial, industrial, and research applications. Here are the key advantages and disadvantages associated with drones equipped with modular payloads:

Advantages

Versatility: Modular payloads allow a single drone platform to perform a wide range of tasks by simply changing the payload. For instance, the same drone can carry a high-resolution camera for photogrammetry on one flight and switch to a thermal camera for inspection or a LiDAR sensor for 3D mapping on another.

Cost-Effectiveness: Investing in a drone capable of carrying different payloads can be more cost-effective than purchasing multiple drones, each dedicated to a specific task. This is especially beneficial for small businesses or researchers with budget constraints.

The DJI Matrice M350RTK is pictured with several of the many sensors available. It is one of the most versatile sUASs on the market today.

Customization for Specific Needs: Modular drones can be customized for specific missions, ensuring that the payload is precisely tailored to the data collection or operational requirements, whether for agricultural monitoring, search and rescue, environmental studies, or construction site inspections.

Ease of Upgrades: Technology evolves rapidly, and having a drone with a modular payload system means new sensors and equipment can be integrated as they become available. This keeps the drone's technology up-to-date without investing in a new drone system.

Reduced Downtime: If a particular sensor or camera needs repair or replacement, it can be easily swapped out without grounding the entire drone system. This is crucial for maintaining operational continuity in commercial applications.

Disadvantages

Initial Cost: Modular systems can cost many times more than a fixed payload drone, sometimes as much as an order of magnitude more.

Drone Cost Projections			
Drone	1	$9,890.00	$9,890.00
Camera	1	$6,250.00	$6,250.00
LiDAR	1	$32,900.00	$32,900.00
Thermal	1	$9,850.00	$9,850.00
Batteries	8	$700.00	$5,600.00
			$64,490.00

A hypothetical example of what a drone can cost

Complexity in Integration: Modular systems can be complex to integrate and manage. Ensuring that the drone's flight control system communicates effectively with various payloads can require additional configuration and technical expertise, which might be a hurdle for some users.

Cost of Multiple Payloads: While the drone might be a one-time investment, high-quality payloads can be expensive. Accumulating several different payloads to cover various applications can result in significant costs.

Increased Maintenance: More components, including various payloads, can increase maintenance challenges. Each payload must be properly maintained and stored when not in use, adding to the operational overhead.

Regulatory and Compliance Challenges: In some jurisdictions, using drones of larger sizes or for different purposes might require additional regulatory compliance, especially when switching between payloads that significantly alter the drone's operation or capabilities, such as switching from a camera to a spray mechanism for agricultural use.

Drones with modular payloads offer great flexibility and efficiency, making them highly valuable for many industries. However, the benefits must be weighed against the increased complexity and potential costs of maintaining multiple high-tech payloads. These factors are crucial in determining whether a modular drone system is a suitable investment for a particular application or project.

Examples:

DJI Matrice M350 RTK

Inspired Flight IF800 Tomcat

Vision Aerial Vector

FreeFly Astro

WISPR Ranger Pro

Arcsky X55

Drones with Fixed Payloads

Drones with fixed payloads are designed for specific tasks. The camera or sensor is permanently affixed and integrated into the drone's architecture. This fixed setup offers streamlined operation and can be optimized for specific applications, ensuring high efficiency and reliability during flights. Here's an exploration of the advantages and disadvantages of drones with fixed payloads, clearly comparing them to their modular counterparts.

Advantages

Simplicity and Reliability: Fixed payload drones are generally simpler to operate due to the lack of interchangeable parts. This simplicity translates to greater reliability, as fewer components could fail and fewer configurations to manage.

DJI Matrice 30t

Optimized Performance: Since the payload is integrated and fixed, it is often optimized for the drone's power and design capabilities, enhancing overall performance. This can result in better flight dynamics, longer battery life, and improved data quality specific to the task for which the drone is made.

Cost-Effectiveness: Fixed payload drones are usually less expensive upfront than modular systems. They provide a cost-effective solution for users who require a drone for a specialized application without needing payload customization.

Ease of Use: The fixed setup eliminates the need for users to understand the technical complexities of payload integration. This makes fixed payload drones particularly appealing to new users and those in industries like real estate or basic aerial photography, where technical demands are less.

Reduced Maintenance: Fixed payload drones often have lower maintenance requirements than modular systems. They have fewer moving parts and less need to detach or swap components.

Disadvantages

Limited Versatility: The biggest drawback of fixed payload drones is their lack of versatility. Each drone is limited to its specific built-in camera or sensor, which cannot be replaced to suit different types of missions or updated as technology advances.

Potential for Obsolescence: As drone technology and sensor capabilities rapidly advance, fixed payload drones can become obsolete more quickly if newer, significantly improved sensors become standard.

Higher Long-Term Costs: While the initial purchase may be cheaper, the inability to adapt the drone for different uses may require purchasing additional drones for different tasks, leading to higher overall costs.

Autel Robotics Evo II Enterprise V3. The Autel Evo II is unique for drones in its size in that it has a changeable payload. It can be adapted to carry either a 512x640 thermal payload or a 6K mapping camera. The connector is not well suited for quick changes and has a fairly low cycle count, but the payload can be changed. The one downfall that many have asked Autel to remedy is that the 6K Sony Camera on the V3 lacks a mechanical shutter, limiting its mapping speed to about 15mph.

Customization Limitations: Users with specific needs that vary slightly from the standard configurations available may find fixed payload drones less accommodating, potentially requiring compromises on data quality or mission goals.

Drones with fixed payloads are best suited for users who need a straightforward, reliable tool for specific applications. They offer ease of use and optimized performance. While they lack the flexibility of modular drones, their lower cost and maintenance requirements make them an attractive option for many scenarios. Understanding the trade-offs between fixed and modular payloads is crucial for individuals and organizations when deciding on a suitable drone to invest in based on their specific operational needs and long-term objectives in drone usage.

Fixed Payload Drones:

DJI Mavic 3 Enterprise

Auzu Robotics Raptor

Autel Robotics 6k Enterprise V3*

DJI Air 2s

Parrot Anafi

* While the Autel Robotics 6k Enterprise V3 is listed as a fixed payload system, it does have a limited ability to switch to the Evo II v3 thermal payload package. This drone is not considered a modular system because the change requires some disassembly with tools, and the ribbon cable connector is subject to a limit on the number of times it can be changed due to accelerated wear.

Perception vs Reality: Drone Performance

Before we examine a few drones that can be employed in survey, mapping, and photogrammetry, there is an 800lb gorilla in the room that needs to be addressed. If you are unfamiliar with drones, you will likely be unfamiliar with this. However, anyone who utilizes this technology must understand that all drone manufacturers list performance under ideal conditions not seen in the real world. Chief among these is range. Take every single range specification with a grain of salt. For example, if an Air 2s says 12km, in real-world conditions, that may only be 2km. Do not assume that larger drones have a much longer range; while the range may be increased, you will not see 15km. You may, under the best of circumstances, see 5km. Even then, the legality of the flight must be taken into consideration. Operations BVLOS (Beyond Visual Line of Site) require an operational waiver. These waivers are difficult to obtain and are typically issued only for a limited duration and location.

Drones Commonly Used in Mapping and Surveying

Fixed Payload Drones

DJI Air 2s

Technical Specifications: The DJI Air 2S, with its compact design and strong imaging capabilities for its compact size, is highly versatile for enthusiasts and professionals. It is equipped with a 1-inch CMOS sensor that captures 20-megapixel photos and 5.4K video, providing exceptional image quality in various lighting conditions. Though it lacks RTK, it is still valuable for many mapping applications where PPK is employed, as well as inspection, and photogrammetry mission

DJI Air 2S

Flight Time: Up to 31 minutes

Range: Up to 12 kilometers

Camera: 1-inch CMOS, 20 Megapixel, capable of capturing 5.4K video at 30fps

Mapping speed: 11-15 MPH

Detailed Description:

The DJI Air 2S is ideal for those needing a budget-friendly beginner drone with decent capabilities and a compact and user-friendly design. Content creators have particularly favored it for its high-resolution camera and ability to shoot 5.4K videos, making it a versatile platform for both mapping and cinematography. The drone's four-directional obstacle sensing makes it safer to operate in complex environments, enhancing its reliability during flight.

The DJI Air 2S also includes features such as Spotlight 2.0, ActiveTrack 4.0, and Point of Interest 3.0, which provide dynamic subject tracking and cinematic video capture, adding to its versatility across multiple use cases. Its compact size makes it highly portable and easy to deploy quickly—ideal for spontaneous shoots in fast-paced environments when data is needed quickly in a compact platform. Many smaller mapping companies still carry one "on the truck" as a reliable yet inexpensive spare.

Autel EVO II Enterprise V3 (6K Mapping Version)

Technical Specifications: The EVO II Enterprise V3 (6K version) is equipped with a 1-inch Sony sensor

Autel Robotics Evo II Enterprise 6k V3 with optional RTK Module

capable of capturing 20-megapixel resolution photos and recording video at 6K quality (5.4k actual). In previous models of the Evo II, that camera was an RYYB camera, which tends to have better color but is much more sensitive to light. In the V3, Autel Robotics made the decision to switch to a Sony camera, likely to increase its credibility in the drone industry. While the Sony camera is very high-quality, like all other drones in its class, it is an RGB camera, causing Autel to move away from its long-time choice of RYYB. The Autel Evo II Enterprise V3 can also be equipped with an optional RTK Module for precision mapping applications. This drone lacks a mechanical shutter, which limits its speed while mapping and reduces its efficiency.

Flight Time: Up to 42 Minutes

Range: Up to 9 kilometers

Camera: 1in CMOS 20 Megapixel

Mapping Speed: 11-15 MPH

Detailed Description: The Autel EVO II Enterprise V3, sometimes dubbed a "business in a box," presents a comprehensive solution for small, full-service drone operations. This drone distinguishes itself through its combination of mapping software, a robust suite for cinematography and photography, its optional RTK module, and a limited ability to change payloads, offering a level of versatility that is virtually non-existent among drones of its size and price range. It is equipped with what Autel dubs "720-degree obstacle avoidance" and is a very capable drone in potentially congested environments. However, this versatility comes with certain compromises. Notably, the absence of a mechanical shutter in the EVO II limits its maximum mapping speed to about 15 miles per hour, significantly affecting efficiency during mapping missions. Despite this limitation, the EVO II Enterprise V3 is more than adequate for small-scale precision mapping and surveying.

DJI Mavic 3 Enterprise

Technical Specifications: Quickly becoming the workhorse of many drone fleets, the DJI Mavic 3 Enterprise is compact and agile, designed for quick deployment and high mobility. It features advanced imaging and data capabilities packed into a portable frame, making it ideal for varied professional applications.

DJI Mavic 3 Enterprise with optional RTK Module

Flight Time: Up to 46 minutes

Range: Up to 15 kilometers

Camera: (2)   Wide – 4/3 CMOS 20 Megapixel Mechanical Shutter (.7s cycle time)

Zoom – 1/2 CMOS 12 Megapixel Electronic/Rolling Shutter

Mapping Speed: 26-31 MPH

Detailed Description: This drone is beneficial for rapid and frequent surveying tasks where ease of use and operational flexibility are paramount. It includes advanced obstacle sensing and avoidance technologies, enhancing its safety for close-range operations. Its mechanical shutter and available RTK Module make it the "go-to" choice for small and medium-scale photogrammetry, mapping, and survey operations. With mapping speeds in excess of 30mph, this drone sets the standard for single payload mapping hardware and is one of the most common mapping drones available today.

Anzu Robotics Raptor

Technical Specifications: The Anzu Robotics Raptor is compact and agile, designed for quick deployment and high mobility. It features advanced imaging and data capabilities packed into a portable frame, making it ideal for varied missions.

Flight Time: Up to 45 minutes

Range: Up to 15 kilometers

Camera: (2)

Anzu Robotics Raptor

Wide – 4/3 CMOS 20 Megapixel Mechanical Shutter (.7s cycle time)

Zoom – 1/2 CMOS 12 Megapixel Electronic Shutter (7x-56x)

Mapping Speed: 26-31 MPH

Detailed Description: This drone is a licensed copy of the DJI Mavic 3 Enterprise manufactured in Malaysia with a completely American-coded and installed software suite. It was brought to market by Anzu Robotics in partnership with Aloft.ai to avoid the "Countering CCP Drones Act" currently before Congress. Because it is a hardware-licensed copy, its software lacks some of the refinement of the Mavic 3 enterprise. However, Aloft.ai, the software partner on the project, is working diligently to remedy that situation and has gone as far as to create a Facebook group of Anzu Robotics Raptor owners to seek input on the highest priorities for the software development team. At the time of its release in May of 2024 the Anzu Robotics Raptor's major benefit was, that it lacked the geofencing common to all DJI drones. However as of late January 2025, all DJI drones have had their geofencing removed and replaced with an FAA accurate warning map.

Modular payload drones

DJI Matrice M350 with the Zenmuse P1 Camera and Zenmuse L2 Sensor

Technical Specifications: The DJI Matrice M350 is designed for robustness and precision in challenging environments. The Zenmuse P1 camera features a full-frame sensor that captures ultra-high-resolution photos, perfect for detailed surveying and mapping. The Zenmuse L2 integrates LiDAR technology with an RGB camera, offering advanced 3D scanning capabilities.

DJI Matrice M350RTK With Zenmuse H20T

Flight Time: Up to 55 minutes

Range: Up to 15 kilometers

Camera: *Zenmuse P1 - 35mm CMOS (Full Frame 24mm x 36mm) 45 Megapixel with three available lenses: 24mm, 35mm (default), and 50mm.

Mapping Speed: Up to 33 MPH

*Note: As with all multi-payload systems, the camera listed is one of several options chosen because it is the default recommended by the manufacturer.

Detailed Description: The Matrice M350 with the Zenmuse P1 is optimized for large-scale photogrammetry, producing detailed maps and models with high-resolution imaging. Adding the Zenmuse L2 sensor allows the same drone platform to perform LiDAR scanning, which is invaluable for generating accurate 3D models and elevation data in close to real-time, even through dense vegetation, enhancing the drone's versatility.

Use Cases:

Topographical Mapping with Zenmuse P1: Excellent for creating detailed elevation models, helpful in planning large-scale development projects.

Construction Monitoring with Zenmuse P1: Provides comprehensive and precise data for overseeing construction progress and ensuring compliance on large sites.

3D Scanning with Zenmuse L2: Ideal for environments where photogrammetry might be hindered by low light or obstacles, such as densely forested areas or urban environments with high-rise structures.

Environmental Analysis with Zenmuse L2: Facilitates the detailed environmental assessment required for impact analysis, allowing planners and ecologists to measure vegetation density, ground cover, and water body volumes accurately.

Additional Insights: The DJI Matrice M350 exemplifies DJI's approach to versatility in its drone platforms. Not only does it support a complete architecture with seamless integration of DJI's own sensors like the Zenmuse P1 and L2, ensuring enhanced interoperability and a unified user experience, but it also offers an open architecture. This flexibility allows for the use of third-party sensors such as the Rock R3 Pro and the Phoenix Recon XT. The open architecture enables users to customize their setups based on specific project needs or to integrate specialized technology that might not be available within DJI's product ecosystem. This dual approach—open for extensive customization and complete for optimized performance—makes the DJI Matrice M350 a highly adaptable platform suitable for a wide range of applications in surveying, construction, and environmental sciences. By accommodating both in-house and third-party solutions, DJI ensures that users can harness the best of drone technology, whether through DJI's advanced systems or external innovations.

Vision Aerial Vector Hexacopter Technical Specifications

Initially designed for the US Secret Service as a heavy-lift, long-endurance drone, the Vision Aerial Vector has six rotors, ultra-powerful motors, and dual battery packs. It features a payload capacity of 5.0 kg (11lb.) and retractable landing gear that deploys quickly, enhancing operational readiness and versatility.

Flight Time: Up to 40+ minutes

Range: Up to 20 kilometers

Payload Capacity: 5.0 kg (11lb.) with versatile mounting points: top, nose, and bottom

Speed: Up to 70 kph (45 mph)

Wind Resistance: Up to 32 kph (20 mph)

Operational Temperature Range: -15ºC to 45ºC (5°F to 113°F)

Vision Aerial Vector Hexacopter

Included Components:

Detailed Description: The Vision Aerial Vector excels in complex, demanding environments. It is designed to carry a range of payloads, including advanced sensors for detailed mapping and surveying tasks. Its quick payload swap capability via the Payload Connection System makes it particularly suitable for rapid deployment in varied missions, ensuring that different sensors and equipment can be utilized effectively and efficiently.

Mapping and Survey Capabilities:

Topographical Mapping: The drone can mount high-resolution cameras and LiDAR systems to create detailed topographical maps and 3D models, which are crucial for planning and development projects.

Geospatial Surveys: The Vector is equipped to handle sophisticated surveying tools, making it perfect for capturing geospatial data essential for environmental and construction applications.

Agricultural Mapping: Supports precision agriculture by carrying multispectral sensors to monitor crop health and manage agricultural inputs efficiently.

Use Cases:

Security Operations: Provides extensive aerial surveillance capabilities necessary for security and monitoring in sensitive or vast areas.

Environmental Monitoring: Adaptable for environmental data collection, crucial for ecological assessments and regulatory compliance.

Industrial Inspections: Ideal for infrastructure inspections, where its stability and endurance can play a pivotal role in maintaining safety and operational continuity.

Research Applications: Facilitates a variety of scientific research endeavors, benefiting from its ability to carry diverse payloads.

Additional Insights: The Vision Aerial Vector merges high payload capacity with extensive flight endurance, featuring a design supporting robustness and adaptability. Its ability to operate under diverse conditions and integrate seamlessly with various mapping and surveying tools makes it an invaluable asset in commercial, governmental, and scientific fields. The dual battery system and advanced flight software enhance its performance, making it a top choice for sophisticated aerial operations.

Freefly Astro with Advanced Payload Options

FreeFly Systems FreeFly Astro

Technical Specifications: The Freefly Astro is built for high-end professional use, emphasizing stability, flexibility, and lifting capacity. This drone is designed to accommodate various payloads, enhancing its utility in diverse operational contexts.

Flight Time: 25-32 min depending on payload

Range: *Not Published – reported 2-12 kilometers*

Camera: *See the List below*

Mapping Speed: *not published – estimated 22-31 MPH*

Detailed Description: The Astro's ability to carry multiple payload options makes it highly versatile and ideal for various applications, from cinematic productions to sophisticated environmental and structural surveys. Its advanced flight capabilities ensure smooth operation, which is crucial for projects requiring precision and high-quality data.

Current Payloads:

Sony A7R IV: Known for its 61 MP sensor, this camera is superb for capturing ultra-high-resolution images, perfect for detailed landscape and architectural photography.

Sentra 65MP Global Shutter: Optimized for high-resolution, rapid capture, suitable for detailed mapping and machine vision applications where motion blur needs to be minimized.

Hovermap ST-X: An advanced LiDAR scanning system that provides detailed 3D mapping capabilities, ideal for environments where GPS is unreliable.

RESEPI Ouster OS1-64: A high-resolution LiDAR sensor that delivers detailed point cloud data for infrastructure and terrain mapping.

GreenValley International X3C-H: Combines LiDAR, photogrammetry, and hyperspectral imaging capabilities, offering a comprehensive survey solution in one package.

Inertial Labs Resepi Teledyne CL-360HD: A cutting-edge LiDAR system that provides high-density point cloud data suitable for detailed topographical surveys and complex engineering projects.

Use Cases:

High-Resolution Mapping with Sony A7R IV: Ideal for capturing detailed aerial photographs for urban planning, landscape architecture, and large-scale surveys.

Precision Agriculture with Sentra 65MP Global Shutter: Allows for detailed crop health analysis, aiding in managing large agricultural operations.

Structural Analysis with Hovermap ST-X: Used in civil engineering to inspect and model structures such as bridges and historical monuments where detailed 3D data is crucial.

Environmental Monitoring with RESEPI Ouster OS1-64: Facilitates the detailed environmental assessment required for impact analysis and conservation planning.

Integrated Surveying with GreenValley International X3C-H: Perfect for projects that require a combination of LiDAR, photogrammetry, and spectral data, such as environmental and forestry studies.

Infrastructure Development with Inertial Labs Resepi Teledyne CL-360HD: Supports complex infrastructure projects by providing detailed LiDAR data for planning and monitoring construction activities.

Additional Insights: The Freefly Astro's adaptability to various high-quality sensors and cameras underscores its role as a leading platform for professional drone operations. Whether it's high-resolution imaging for cinematic productions or detailed environmental surveys, the Astro allows operators to select the best tools for their specific needs, making it a valuable asset for professionals across industries.

Inspired Flight IF800 Tomcat with Advanced Payload Options

Technical Specifications: The Inspired Flight IF800 is designed for versatility and robust performance, capable of carrying a variety of high-end payloads. This includes the Phase One iXM-100, Sony a7R IV, Sony ILX-LR1, and the Sentra 65R, each offering unique capabilities tailored to specific surveying and imaging needs.

Inspired Flight IF800 Tomcat

Flight Time: Up to 54 minutes

Flight time Real World, with Payload: up to 40 Minutes

Ideal Range: Depends on GCS chosen, can range from 9km and up.

Real World Range: Based on GCS Chosen, Herelink Blue, a common choice, can provide more than 3.5km

Camera: *See List below*

Max speed: 49mph

Mapping Speed: *not published – estimated 26-33 MPH*

Wind Resistance: Level 6 *(23kts)*

Detailed Description: The IF800 is built to support complex missions in varied environments. Its focus on durability and payload flexibility allows it to accommodate high-resolution cameras and specialized sensors, making it suitable for a broad range of applications, from precision agriculture to advanced scientific research.

Current Payloads:

Phase One iXM-100 The Phase One iXM-100 is a sophisticated metric aerial camera boasting a 100MP backside-illuminated medium-format sensor that enhances light sensitivity and dynamic range, making it ideal for UAV-based imaging missions. It supports a range of high-resolution RSM lenses from 35mm to 300mm, ensuring superb image sharpness across various flight conditions. This camera is particularly noted for its fast response, robust build, and high productivity, suited for detailed mapping, surveying, and inspection tasks. With features like a 3fps capture rate and high-resolution capabilities, the iXM-100 is designed to deliver superior quality aerial imaging and flexible operation across diverse applications.

Sony a7R IV: The Sony a7R IV is a high-resolution full-frame mirrorless camera designed to deliver exceptional image quality and versatile performance, making it ideal for photogrammetry and advanced aerial imaging. It features a 61MP Exmor R CMOS sensor, which provides striking detail and resolution, beneficial for detailed mapping and survey applications. The camera is known for its robust autofocus system, offering comprehensive coverage and fast response, which is crucial for capturing sharp drone images. Its dynamic range and high ISO capabilities also ensure excellent performance in various lighting conditions, making it a solid choice for professional drone-based photography and videography.

Sony ILX-LR1 The Sony ILX-LR1 is a highly specialized full-frame camera explicitly engineered for industrial applications, including drone operations. It features a 61MP Exmor R CMOS sensor and is significantly lighter than typical cameras, weighing only about 0.54 lbs. This makes it exceptionally suitable for extended drone flight times and complex aerial tasks. The ILX-LR1 lacks traditional camera components like an LCD monitor or viewfinder, optimizing it for remote operations via Sony's SDK. It supports UHD 4K60 video recording and has advanced connectivity options such as locking USB-C and Micro-HDMI outputs, making it ideal for high-resolution aerial photography and video capture in industrial settings.

Sentera 65R: The Sentera 65R is a highly advanced ultra-high-resolution aerial sensor specifically designed for integration with drone systems to enhance the precision and efficiency of aerial imagery. It achieves a ground sampling distance of 0.45 cm per pixel at 125 feet and 1.43 cm per pixel at 200 feet, making it exceptionally capable of detailed surveying and mapping tasks. With a global shutter and a robust design tailored for drone use, the 65R excels in capturing clear and precise images, supporting various applications such as agricultural analytics and environmental monitoring.

The NextCore Lumos 100XT: The Lumos 100XT is a high-performance UAV LiDAR system compatible with the Inspired Flight IF800 Tomcat drone. This system is renowned for its reliability and ease of use, making it ideal for extensive aerial surveying and mapping tasks. It features automatic in-flight INS calibration and user-friendly software, significantly simplifying the workflow, allowing for efficient field operations. The Lumos 100XT can generate high-density point clouds with a range of up to 120 meters, making it particularly effective in rugged and vegetated terrains.

Future Developments and Compatibility: Inspired Flight is actively working towards developing its integrated architecture similar to the DJI Matrice M350, aiming to enhance interoperability and streamline operations across its drone fleet. Meanwhile, the platform maintains an open architecture approach, allowing the integration of third-party sensors like the Rock R3Pro and the Phoenix ReconXT.

This flexibility ensures that the IF800 remains a highly adaptable solution, capable of meeting diverse operational demands by incorporating both proprietary and external technological advancements.

Additional Insights: Inspired Flight's strategic development of its own architecture signals its commitment to providing a unified and enhanced user experience while retaining the versatility offered by compatibility with a wide range of third-party sensors. This dual approach caters to both specific customer needs and broader industry requirements, making the IF800 ideal for professionals looking for a high-performance, adaptable drone platform.

Fixed Wing Drones

Atmos Marlyn

The Marlyn drone is a technologically advanced unmanned aerial vehicle (UAV) designed by Atmos UAV, tailored explicitly for mapping and surveying purposes. It is notable for its VTOL (Vertical Takeoff and Landing) capabilities, combining a multirotor's flexibility with a fixed-wing aircraft's efficiency and speed. This hybrid design allows the Marlyn drone to take off and land in confined spaces while covering large areas quickly during flight.

Atmos Marlyn

Key Specifications:

Weight: Approximately 4 kg (including payload).

Wingspan: 1.7 meters, providing a balance of aerodynamic efficiency and compactness.

Payload Capacity: Equipped to handle various sensors, with a standard high-resolution RGB camera and options for multispectral and thermal sensors.

Flight Time: Up to 60 minutes on a single battery charge, enabling it to cover areas of up to 150 hectares per flight.

Cruising Speed: 45-70 km/h, adjustable based on the mission requirements.

Maximum Altitude: Capable of flying up to 1,200 meters above sea level but typically restricted by local aviation regulations.

Operating Temperature: -10°C to +40°C, designed to function in various environmental conditions.

Features:

Advanced Navigation Systems: Includes GPS and GNSS capabilities for precise geolocation. They are integrated with RTK (Real-Time Kinematic) for centimeter-level accuracy in mapping.

Robust Software Integration: Compatible with a variety of GIS and photogrammetry software, facilitating easy integration of the data collected into different platforms and workflows.

Automated Flight Planning: Users can pre-plan missions with specific waypoints, altitudes, and other parameters, which the drone will autonomously follow once airborne.

Safety Features: Equipped with emergency procedures such as automatic return-to-home on low battery or lost communication.

Durable Design: Built to withstand challenging weather conditions, with a robust body that minimizes dust and light rain damage.

The Marlyn drone is an excellent tool for topography, agriculture, forestry, mining, and environmental monitoring professionals. It provides high-resolution aerial data that can significantly enhance the accuracy and efficiency of spatial analysis and decision-making.

Quantum Systems Trinity F90+

The Quantum Systems Trinity F90+ is a cutting-edge fixed-wing UAV designed for professional use in geospatial, agricultural, and surveillance industries. It stands out with its unique combination of VTOL (Vertical Takeoff and Landing) capabilities and long-endurance fixed-wing flight, making it ideal for high-precision mapping and data collection over large areas.

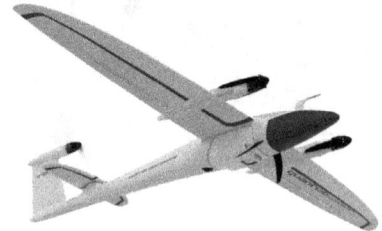

Trinity F90+

Key Specifications:

Weight: 2.5 kg (5.5 lbs), making it lightweight and easy to transport.

Wingspan: 2.3 meters, which helps to achieve stability and efficiency in flight.

Payload Capacity: Capable of carrying up to 700 grams of payload, allowing for the integration of various sensors and cameras.

Flight Time: Up to 90 minutes on a single battery charge, depending on payload and flight conditions, which is excellent for extended operations over large territories.

Cruise Speed: Typically operates at 36MPH (60 km/h), offering a good balance between speed and power efficiency.

Maximum Range: Up to 60 Miles (100 kilometers), subject to regulatory restrictions and operational settings.

Operating Temperature: Designed to operate within -10°C to 50°C, accommodating a wide range of environmental conditions.

Features:

Advanced Navigation and Control: Equipped with GNSS (Global Navigation Satellite System) and optional RTK (Real-Time Kinematic) for precise navigation and positioning with centimeter-level accuracy.

Payload Flexibility: Supports a variety of payloads including RGB, multispectral, thermal, and LiDAR sensors, adaptable via its modular payload bay.

Automated Flight and Mission Planning: Features an intuitive mission planning software that allows for easy design of flight paths, automatic takeoff, flight execution, and landing.

Safety and Reliability: Includes fail-safe mechanisms such as automatic return-to-home on loss of communication or low battery, ensuring safety during missions.

Durable and Robust Design: The Trinity F90+ is built to withstand harsh conditions. Its design resists dust and light precipitation, ensuring reliability in diverse operational environments.

The Quantum Systems Trinity F90+ is renowned for its operational efficiency and accuracy, making it a top choice for professionals who require reliable and precise aerial data collection capabilities. Its versatile design accommodates a broad range of industry applications, from environmental monitoring to precision agriculture and urban planning, providing comprehensive insights and enhancing decision-making processes.

The WingtraOne GEN II

The WingtraOne GEN II is an advanced fixed-wing vertical takeoff and landing (VTOL) drone engineered explicitly for high-precision aerial surveying and mapping across various industries such as agriculture, mining, construction, and environmental monitoring. This second-generation model by Wingtra continues to push the boundaries of UAV technology with enhanced features and improved operational efficiencies.

Key Specifications:

Weight: Approximately 3.7 kg, designed for easy transport and setup.

Wingspan: 1.25 meters, providing excellent aerodynamic performance and stability in flight.

Payload Capacity: Capable of carrying high-resolution cameras and sensors tailored to different mapping needs.

Flight Time: Up to 59 minutes, enabling extensive coverage of up to 400 hectares (990 acres) in a single flight at the lowest altitude.

Cruise Speed: 16 m/s (57 km/h), ideal for efficient mapping operations over large areas.

Maximum Altitude: Flight capable of up to 5000 m above sea level, which is beneficial for high-altitude survey missions.

Operating Temperature: Operates effectively in temperatures ranging from -10°C to 40°C, allowing deployment in diverse climates.

Features:

Advanced Imaging Capabilities: The WingtraOne GEN II supports a variety of payloads, including the 42 MP Sony RX1R II camera for RGB imaging and specialized sensors for multispectral and thermal imaging. This versatility makes it suitable for detailed geological surveys, precision agriculture, and environmental research.

Robust Navigation Systems: Equipped with PPK (Post-Processed Kinematic) as standard, enhancing the geolocation precision to centimeter-level accuracy without the need for ground control points in many cases.

Efficient and Safe Operations: The VTOL technology allows it to take off and land vertically in confined spaces, eliminating the need for a runway and reducing the risk of landing damage associated with traditional belly-landing fixed-wing drones.

User-friendly Interface: WingtraPilot software provides an intuitive planning and operation platform, making it easy to pre-plan missions, monitor them in real time, and efficiently manage the collected data.

Durable Construction: The drone is built with rugged materials to withstand the demands of regular field use and varying environmental conditions.

The WingtraOne GEN II stands out for its exceptional data accuracy, ease of use, and versatile payload options, making it an ideal solution for professionals seeking reliable and precise geospatial data across a wide range of applications.

What's next: Choosing a suitable sensor for a drone mapping project depends on the specific requirements of the project. High-resolution cameras like the Zenmuse P1 or Sony A7R IV are suited for projects requiring detailed photogrammetry. At the same time, specialized sensors like multispectral, thermal, and LiDAR cater to more complex needs that require these specialized tools. Understanding these capabilities allows drone operators to tailor their equipment choices to the project's demands, ensuring that the data collected meets the precision and accuracy required for professional-grade survey, mapping, and analysis. As technology advances, the integration and capabilities of these cameras and sensors continue to evolve, pushing the boundaries of what can be achieved in drone mapping. Over the following two chapters, we will explore the technology behind sensors, discuss their differences, and give you some insight into what to choose for your needs.

Chapter 3: Cameras for Drones

Drone mapping technology is entirely defined by the sensors utilized. Each sensor type serves a specific purpose tailored to various mapping needs. The success of a drone mapping project heavily relies on the quality and type of cameras and sensors employed. These tools are not mere accessories but an integral part of capturing the detailed, accurate data necessary for professional-grade maps and models. This chapter expands on the types and functions of cameras typically utilized in drone mapping, focusing on their functionalities and specific applications. While there are many types of sensors used by drones, most professional pilots will agree, the camera is the most used tool of them all.

In the following section, we will delve, ever so briefly, into why these sensors are so important. By exploring the basics of photography and photogrammetry, the reader should better understand the need for high quality equipment for these missions.

More than 80% of all drone mapping operations are performed with "RGB" or Red, Green, Blue Photographic cameras, so it's essential to begin with the basics of photogrammetry and how it differs from standard photography.

Photography:

This is not a book about photography. To be clear, a single work on photography may make up many volumes and thousands of pages. I have neither the knowledge nor the time to write a book about the details of all of photography. That being said, it is important to understand a few basic fundamentals as they apply to using drones in photography, and by extension photogrammetry. So, we'll discuss the basics of photography and how it relates to photogrammetry. Please understand this is not a complete work and even some fundamental basics may be left out if they do not directly apply to photography as it relates to photogrammetry in drones.

The Exposure Triangle:

At the heart of photography lies the "Exposure Triangle", essential across all types of photography, whether it's using the latest iPhone or a $55,000 Phase One XF IQ4. The exposure triangle is about managing light's entry into the camera. It consists of three main elements—aperture, shutter speed, and ISO—each play a critical role in light manipulation. Properly balancing these elements can yield remarkable results, even with average sensors.

The exposure triangle is a fundamental concept in photography that describes the relationship between three crucial elements of exposure: aperture, shutter speed, and ISO. Understanding and balancing these three elements allow photographers to control the brightness of their photos, depth of field, motion blur, and noise levels. Here's a quick look at each component:

Aperture

This refers to the size of the lens opening through which light enters. A wider aperture (a lower f-number, such as f/2.8) allows more light to pass through to the sensor, which is great for low light conditions and also creates a shallow

depth of field (blurry background). A smaller aperture (a higher f-number, like f/16) lets in less light but increases the depth of field, making more of the image appear sharp.

Shutter Speed

This is the duration for which the camera's shutter is open to expose light onto the camera sensor. Faster shutter speeds (like 1/1000 second) allow less light to hit the sensor but are excellent for freezing motion. Slower shutter speeds (such as 1 second) expose the sensor to more light and can capture the effect of motion, such as blurring moving water or lights

ISO

This measures the sensitivity of the camera's sensor to light. Lower ISO values (like 100 or 200) are used in bright conditions to avoid overexposure and maintain the highest image quality, free from noise. Higher ISO values (like 3200 or 6400) are necessary in low light situations but can result in more noise or grain in the image.

Balancing the Triangle

Understanding the exposure triangle involves learning how to balance these three settings to achieve a desired photographic effect while maintaining proper exposure. Changing one element will often require adjustments to the others. For example, if you increase the aperture size, you might need to decrease the ISO or increase the shutter speed to prevent overexposure.

Effective use of the exposure triangle allows photographers to exert creative control over their images, whether they're aiming for sharply detailed, brightly lit scenes or moody, motion-blurred shots with a focus on light play. Mastery of these elements also helps in adapting to various lighting conditions and subjects.

Photogrammetry in Drone Mapping

Photogrammetry is a transformative technique that uses software to assemble photographs into one cohesive high resolution image or "map" to measure and interpret the features of the Earth's surface. It is crucial in turning simple photos into valuable digital meshed images that can be analyzed and utilized across various sectors. The following sections explore the basics of photogrammetry process from planning missions to data capture, and data processing, and its diverse applications in industries such as survey, civil engineering, construction, agriculture, archaeology, and inspection.

Basics of Photogrammetry

Photogrammetry involves using photographic images to create maps, detailed drawings, or 3D models of real-world objects or scenes. It harnesses the power of perspective, where two photos of the same area, taken from slightly different angles, produce a three-dimensional visualization of the area.

When applied to drone technology, photogrammetry allows for capturing high-resolution images over vast areas of land or inaccessible terrain quickly and efficiently. Drones equipped with high-resolution cameras fly over the area of interest, capturing overlapping photographs at strategic angles and elevations. These overlapping images ensure that every landscape feature is captured from multiple viewpoints, increasing the depth of the data collected.

In general, the effectiveness of drone photogrammetry hinges on several factors:

Camera Quality: Higher resolution cameras with larger sensors yield more detailed images, allowing for more precise measurements and clearer models. However, its important to understand that, sensor type and quality matter as much as pixel count.

Flight Altitude and Speed: These affect image clarity and the area covered in each photograph. Lower altitudes allow for greater detail but cover less ground. Higher altitudes allow for greater coverage much faster, but may sacrifice some detail in the process.

The measure of resolution in drone photogrammetry is commonly referred to as GSD, or Ground Sampling Distance. Ground Sampling Distance is generally measured in centimeters. 1 GDS equates to 1 centimeter from the center of 1 pixel of an image to the center of any adjacent pixel. So, a map where the image is 1cm between pixel centers is 1GSD. GSD can be controlled in one of two ways, either by altitude or lens/sensor size. We will delve further into this later in this chapter but suffice it to say, GSD as it equates to drone mapping is one of the most critical factors in determining the quality of orthomosaic images.

Overlap of Images: When flying mapping or orthomosaic missions of any type, flying an overlap between images is crucial to generate a quality orthomosaic image. Generally, a minimum of 65% by 75% overlap is necessary for mapping, and up to 83% by 83% is necessary for accurate 3D modeling. The flip side to this is, too much overlap can increase the size of the dataset to unprocessible levels while also reducing the amount of points available for alignment.

Lighting Conditions: Adequate natural lighting is crucial for clear, usable photographs. It is best to begin dataset captures no earlier than 90 minutes after sunrise and end dataset captures about 90 minutes before sunset. Sufficient lighting is even more crucial for high-resolution mapping as there can be some light loss in color balancing while the dataset is being assembled.

Process of a Photogrammetry mission

The process of drone photogrammetry will be covered in more detail in later chapters, however, here's a brief break down of its several steps:

Flight Planning: This initial stage involves defining the area to be surveyed and planning the drone's flight path. Factors such as the size of the area, the topography, and the purpose of the survey dictate the flight pattern, altitude, and image overlap required. In standard mapping scenarios, one of two main types of flight will be performed: a down-and-back pattern or a cross-hatch pattern. Down and back will likely be the pattern used on standard nadir mapping or "smart oblique" 3D missions. For standard 3D missions, a cross-hatch or standard oblique pattern will be flown.

Image Capture: During the flight, the drone automatically captures images based on the predetermined flight plan. This is where a mechanical shutter vs an electronic shutter will come into play. Most flight planning software can now take the data regarding GSD and overlap and set the speed and image capture interval automatically. GPS data is often recorded simultaneously to geo-tag each photograph for precise location referencing. If a drone is equipped with integrated RTK, the data inserted into the photo or the geo-tag will contain highly accurate data, generally down to 1cm+1ppm.

Image Processing: After the flight, the collected images are processed using specialized software such as Pix4D, Agisoft Metashape, or Autodesk Recap. This software aligns images, corrects any distortions, and stitches them together based on their overlapping areas and GPS tags. The result is a detailed, accurate photomosaic and a 3D digital surface model (DSM) or digital elevation model (DEM).

Analysis and Interpretation: The final models and maps are analyzed to extract meaningful information. This step may involve additional software tools depending on the application, ranging from simple measurement tools to complex analytical software for advanced interpretations.

Applications of Photogrammetry, Photogrammetry's versatility makes it applicable in numerous fields:

3D Renderings: Photogrammetry can generate high-quality 3D models from 2D images, useful for virtual reality, gaming, and architectural visualization.

Agriculture: Farmers use drone photogrammetry for crop monitoring, which involves assessing plant health, estimating yield, and planning irrigation systems. The precise data helps in making informed decisions that enhance productivity and sustainability.

Archaeology: Photogrammetry provides a non-invasive means of exploring and documenting historical sites. It allows for creating detailed models of dig sites, artifacts, and landscapes, which helps preserve and study cultural heritage.

Construction: In the construction industry, photogrammetry is used to monitor project progress, manage assets, and maintain documentation. It enables project managers to track changes over time, inspect hard-to-reach areas, and ensure that construction meets planned specifications.

Mapping and Survey: Photogrammetry is crucial for creating accurate maps and surveying land. It's used to update geographic information systems (GIS), plan urban development, and analyze terrain for various engineering projects.

Inspection: For infrastructure like bridges, dams, or pipelines, photogrammetry allows for detailed inspection without the need for scaffolding or downtime. It helps detect defects, measure wear, and plan maintenance or repairs efficiently.

Mining: In mining, photogrammetry aids in volume calculations of stockpiles, safety monitoring of pit walls, and mapping new excavation areas.

Safety: Photogrammetry enhances industrial safety by enabling remote inspection of hazardous areas, reducing human exposure to dangerous environments.

Environmental Monitoring: It's used to track environmental changes, monitor deforestation, assess industrial impacts, and ensure compliance with environmental regulations.

Film and Entertainment: The film industry leverages photogrammetry for creating realistic digital sets, props, and characters for CGI, improving visual effects and reducing production costs.

Forensic Science: In crime scene investigation, photogrammetry reconstructs scenes in 3D for evidence analysis, documentation, and courtroom presentations.

Real Estate: Real estate uses photogrammetry for virtual property tours, accurate property visualizations, and marketing, especially for complex structures.

Cultural Heritage: For cultural preservation, photogrammetry creates detailed digital replicas of historical sites and artifacts, aiding restoration and education.

Civil Engineering: Photogrammetry in civil engineering helps monitor structural health, inspect bridges and tunnels, analyze land deformation, and manage infrastructure projects.

Disaster Response: Post-disaster, photogrammetry quickly maps affected areas for damage assessment, planning relief, and documenting changes for recovery efforts.

Photogrammetry has revolutionized how data is collected and utilized across various sectors. Its integration with drone technology has enhanced its capabilities, making it a powerful tool for comprehensive, efficient, and cost-effective data collection. As drone and camera technologies continue to advance, the scope and accuracy of photogrammetry will only improve, broadening its applications and making its benefits even more significant.

Photographic Cameras

It is important to note that, as with so many technologies adapted for use with drones, digital cameras have existed for decades, and literally dozens of technologies have evolved around their development and use. It is impossible to cover every aspect of the function of modern digital cameras in the span of this text; volumes could easily be written about any of the many different technologies employed. That being the case, this section will focus solely on the information needed to select the hardware to fit the reader's needs should you choose to enter the field of drone mapping. No matter what camera payload you buy, there will be compromises, accept that. From the 3200-megapixel LSST at the Vera C. Rubin space observatory to the camera in your iPhone 15 pro, cameras are compromises.

Drone Mapping: It's All About the Sensor

When it comes to drone photogrammetry, the choice of camera or sensor is paramount. The quality of data collected depends heavily on the sensor's capability. Understanding why specific cameras are preferred, what makes a good mapping camera, and why some should be avoided is crucial for successful aerial photogrammetry, surveying, or mapping.

Importance of Camera Choice: High-quality sensors capture finer details essential for creating accurate maps and models. Cameras used in photogrammetry typically feature a large CMOS sensor with large pixels, capturing more light and providing greater dynamic range, which enhances image clarity under various lighting conditions. Additionally, certain cameras, such as the 45-megapixel Zenmuse P1 and the 20-megapixel 4/3 CMOS on the Mavic 3 Enterprise and Multispectral drones, are equipped with technology that "time syncs" the GPS data with RTK accuracy to the image. RTK would be effectively useless for drone mapping without this sync data. Other camera/drone combinations such as the Sony ILX-LR1/Grimsey gimble/various drone companies combos, may generate a separate file, which can then be paired to the images in the post flight processing software, adding an additional step to image processing.

Rolling Shutter vs Mechanical Shutter vs Global Shutter

In this section, we discuss camera shutters; there are few places in photography where shutter choice is as vital as it is in photogrammetry, so this section will explore the sometimes very confusing world of camera shutters.

Rolling Shutter effect on image if the camera is in motion

Rolling (Electronic) Shutter

Function and Limitations: An electronic shutter does not involve any physical movement but electronically controls the activation of the sensor pixels. The sensor's pixels are turned on and off in sequence, typically from top to bottom, which can lead to discrepancies in the timing of data capture across the image when the camera is in motion.

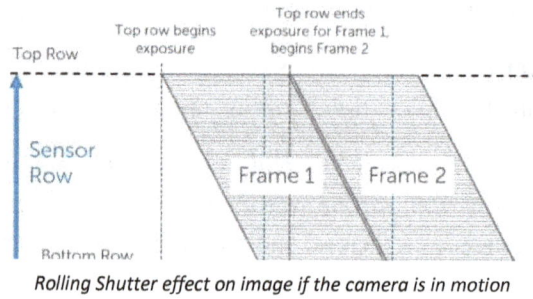

Advantages:

Smaller: Electronic shutters allow for a much smaller, compact design. If your smartphone has a camera, it is equipped with an electronic shutter. In the early days of consumer drones in the early 2000s, virtually all were equipped with rolling shutters as a weight-saving measure.

No Moving Parts: Unlike a mechanical shutter, which is exactly as it sounds, mechanical, a rolling shutter has no moving parts. Every aspect of the photo "transaction" is done electronically, which contributes dramatically to the overall size reduction.

Disadvantages:

Rolling Shutter Effect: As drones move, objects captured by different areas of the sensor at different times can appear skewed or wavy. This is particularly problematic when capturing high-speed images or when the drone is turning. It is commonly assumed that a faster

Camera	
Camera maker	Autel Robotics
Camera model	XL720
F-stop	f/2.8
Exposure time	1/15 sec.
ISO speed	ISO-198
Exposure bias	0 step
Focal length	11 mm
Max aperture	2.97085360437631607
Metering mode	Center Weighted Average
Subject distance	
Flash mode	No flash
Flash energy	
35mm focal length	29

shutter speed will reduce or resolve this issue, sadly, that is not the case. Shutter speed controls how long a pixel is exposed to light, not how long it takes for the shutter to cycle through all the pixels. The shutter transaction typically begins at the top of an image and progresses to the bottom; regardless of how long each pixel is exposed, the process can take up to 1/10th of a second, depending on the make and model of the camera. Compare that to a mechanical shutter, where every pixel is exposed simultaneously in as

Camera	
Camera maker	Anzu
Camera model	Raptor
F-stop	f/2.8
Exposure time	1/725 sec.
ISO speed	ISO-100
Exposure bias	0 step
Focal length	12 mm
Max aperture	2.97
Metering mode	Average
Subject distance	0 mm
Flash mode	No flash
Flash energy	
35mm focal length	24

little as 1/2000th of a second, and it becomes much easier to understand why the attributes of a mechanical shutter make it the better choice.

Reduced Accuracy: For mapping purposes, these distortions can compromise the geometric fidelity of the images, making them less suitable for creating precise maps and models.

Mechanical Shutter

In general, mechanical shutters have several advantages and a few disadvantages.

Advantages:

No Rolling (Electronic) Shutter Effect: Mechanical shutters eliminate the rolling shutter effect, which is critical for aerial mapping where the drone is constantly in motion. This ensures that buildings, vehicles, trees, and other structures are accurately represented in the photographs without distortion.

High Image Quality: The instantaneous capture of the image frame reduces blur and provides higher image fidelity, which is crucial for detailed and accurate photogrammetric processing.

Faster Mapping Speeds: Because mechanical shutters do not suffer from the image drift of electronic rolling shutters, drones equipped with mechanical shutters can map at much faster speeds than those equipped with electronic rolling shutters.

Disadvantages:

Larger: Because mechanical shutters are equipped with a moving "curtain," they are larger, typically 4 to 5 times the size of a small rolling shutter camera. This increased size adds both weight and complexity to the camera.

Mechanical: Mechanical means just what it says. These cameras have moving parts, and as Elon Musk said at SpaceX, "The best part is no part."

Function and Benefits: A mechanical shutter operates similarly to a curtain in front of the camera sensor, which opens to begin capturing an image and closes to end the exposure. The entire sensor is exposed at once, capturing the entire image simultaneously. This global exposure prevents or dramatically reduces distortions in the image caused by moving objects or the drone's motion during image capture. The faster our mechanical shutter functions the faster it can capture data and the faster a drone can fly to map. It's important to note that there are three main types of mechanical shutters, they are:

Leaf Shutter: This type of shutter consists of several overlapping metal blades, much like the leaves of a plant. When the shutter is triggered, these blades move to expose the camera's sensor or film. Leaf shutters are known for their quiet operation and suitability for high-speed flash synchronization.

Focal Plane Shutter: Unlike leaf shutters, focal plane shutters are located near the camera's focal plane, typically just in front of the film or image sensor. These shutters consist of two curtains that move across the film or sensor plane, with the second curtain following the first to create the exposure. Focal plane shutters are common in SLR (Single-Lens Reflex) and DSLR (Digital Single-Lens Reflex) cameras.

Central Shutter: Also known as a "diaphragm shutter," a central shutter is located within the lens assembly itself. It consists of blades that open and close like an aperture to control the duration of exposure. Central shutters are often found in medium format and large format cameras.

In drone operations, it is not uncommon to see another shutter commonly referred to as a central leaf shutter. A central leaf shutter, also known as a hybrid shutter, combines the features of both central and leaf shutters into a single mechanism. This type of shutter is typically found in specialized lenses designed for specific camera systems, particularly medium-format cameras.

Here is how a central leaf shutter typically works:

Location: A central leaf shutter is located within the lens assembly, near the aperture diaphragm, much like a central shutter.

Design: The central leaf shutter incorporates elements of both central and leaf shutter designs. It consists of multiple blades arranged in a circular pattern, similar to the blades of a leaf shutter. However, unlike a traditional leaf shutter, these blades open and close from the center of the lens outward, covering the entire aperture area.

Operation: When the shutter is triggered, the blades of the central leaf shutter open from the center outward to expose the camera's sensor or film. Once the exposure duration is complete, the blades close back to the center, ending the exposure. This design allows the shutter to control the exposure duration while maintaining a compact size within the lens assembly.

Advantages: Central leaf shutters offer several advantages:

- They provide precise control over exposure duration, similar to central shutters.

- They offer the quiet operation and low vibration associated with leaf shutters.

- They can synchronize with flash at high speeds, like leaf shutters, allowing for creative lighting techniques and better control over ambient light.

- They are typically compact and lightweight, making them suitable for use in portable medium-format camera systems like drones.

Global (Electronic) Shutter

Function and Benefits: Similar to a mechanical shutter in its ability to capture an entire image simultaneously, A global shutter captures the entire image on the sensor by exposing all pixels at the same exact moment. This method contrasts with rolling shutters, which expose different parts of the image at slightly different times, and with mechanical shutters because, like rolling shutters, Global shutters are completely electronic and have no moving parts. Global Shutters are particularly beneficial in fast-moving scenarios, such as aerial surveys conducted from drones, because they eliminate distortions caused by motion, both of the subject and the camera itself.

Advantages:

No Rolling Shutter Effects: Global shutters are ideal for capturing images where high-speed or rapid motion is involved. They prevent distortions such as skewing, smearing, or wobbling of the image, which are common with rolling shutters. This is crucial in high-speed photography or when capturing fast-moving subjects from moving platforms like drones.

Uniform Exposure: Since every pixel on the sensor captures light simultaneously, images have uniform exposure and timing, which is vital for consistent lighting and color in fast exposures.

Better for High-Speed Events: Global shutters are better suited for capturing events in microseconds, such as scientific experiments or explosions, where capturing the moment precisely as it occurs is critical.

Disadvantages:

Reduced Dynamic Range: Global shutters often suffer from a slightly reduced dynamic range compared to rolling shutters and mechanical shutters. This is due to the architecture of the sensor, which must accommodate additional circuitry for each pixel to enable simultaneous capture.

Increased Cost and Complexity: Cameras equipped with global shutters are more expensive and complex to manufacture. The technology required to enable all pixels to read out simultaneously can lead to higher production costs.

Increased Power Consumption: Global shutter sensors consume more power than both rolling shutters and mechanical shutters. This increased power draw can be a concern in battery-operated devices like drones, where power efficiency is essential.

Global shutters provide significant advantages in applications where capturing motion accurately and without distortion is crucial. They are particularly valuable in aerial photogrammetry, fast-action sports, and scientific research, where the exact synchronization of the image capture with the event is necessary. However, these benefits come at the cost of increased system complexity, power consumption, and potentially higher financial investment.

Global Mechanical Shutters:

Function and Benefits: Global mechanical shutters combine the advantages of both global shutters and mechanical shutters. Like global shutters, they capture the entire image on the sensor simultaneously, ensuring uniform exposure across the frame. Additionally, they employ the physical barrier mechanism of mechanical shutters, which eliminates the rolling shutter effect and prevents distortions caused by motion, both of the subject and the camera itself.

Advantages:

Distortion-Free Imaging: Global mechanical shutters ensure distortion-free imaging, making them ideal for scenarios where capturing motion accurately and without distortion is critical. This is particularly advantageous in applications such as aerial photography, where drones are in constant motion.

High Image Quality: The simultaneous capture of the entire image frame results in high image fidelity with minimal blur, enabling detailed and accurate image processing. This is essential for applications requiring precise photogrammetry or scientific research.

Versatility: Global mechanical shutters offer versatility in capturing a wide range of scenes, from high-speed events to static subjects, without compromising image quality or introducing distortions.

Reduced Power Consumption: Global mechanical shutters can offer improved power efficiency compared to purely electronic shutters. While they consume more power than purely mechanical shutters due to their electronic components, the simultaneous exposure reduces the overall exposure time, leading to energy savings.

Disadvantages:

Increased Complexity and Cost: Global mechanical shutters are more complex and costly to manufacture than purely mechanical or purely electronic shutters. This is because they require both mechanical components for the shutter mechanism and electronic components for simultaneous exposure of the sensor.

Potential Weight and Size Increase: Incorporating both mechanical and electronic components may increase size and weight compared to simpler shutter designs. This added bulk can be a consideration, particularly in applications where size and weight are critical factors.

Maintenance Requirements: Like any mechanical component, global mechanical shutters may require regular maintenance to ensure proper functioning over time. This can add to the overall cost of ownership and may necessitate periodic servicing.

Global mechanical shutters offer a compelling solution for applications requiring distortion-free imaging and precise capture of fast-moving subjects. In the best of scenarios, a camera like the Zenmuse P1 comes equipped with a Global Central Leaf Mechanical Shutter can provide unique, fully integrated mapping solutions. While they may come with increased complexity, cost, and potential maintenance requirements, their ability to combine the advantages of global and mechanical shutters make them well-suited for a variety of demanding photogrammetry and imaging tasks.

In the realm of camera technologies, the differences between rolling shutters, global shutters, mechanical shutters, and global mechanical shutters significantly influence the quality and applicability of captured images. Rolling shutters are cost-effective and efficient for standard video capture, but they can introduce distortions in fast-moving scenarios due to the sequential exposure of image sensors. Global shutters, on the other hand, capture the entire image simultaneously, eliminating motion distortions and making them ideal for high-speed photography and video. Mechanical shutters feature physical components that control exposure, provide high image fidelity and precise control, and are suitable for professional photography under varying lighting conditions. Global mechanical shutters combine the benefits of both mechanical precision and global coverage, ensuring distortion-free images with precise exposure control, thereby offering an advanced solution for applications demanding the highest level of accuracy and speed, such as scientific research and high-speed aerial mapping. Each type of shutter has its own strengths and limitations, making it suitable for different photographic needs and specializations.

Pixel Count in Drone Photography and Mapping: Is More Always Better?

Pixel count, typically measured in megapixels (MP), is a fundamental specification in drone photography and mapping. It directly affects the resolution and detail of the imagery captured. Pixel count refers to the total number of pixels in an image, calculated by multiplying the number of horizontal pixels by the number of vertical pixels in the sensor's array. For instance, a camera with a resolution of 4000 x 3000 pixels has a pixel count of 12 million pixels or 12 megapixels. Each pixel captures a part of the image, influencing the potential level of detail in the final image.

High pixel counts are crucial for capturing finer details, which is essential in drone-based photogrammetry or surveying where precise measurements are needed. A higher pixel count allows for greater detail and clarity when zooming into images, ensuring that the quality does not degrade. In applications like mapping and surveying, this translates into more accurate outputs such as orthomosaics or 3D models. These high-resolution images provide more data points, which are particularly useful in sectors like agriculture for monitoring crop health, construction for tracking site progress, and environmental studies for assessing natural resources or disaster impacts.

However, high pixel counts come with their challenges. Larger image files result from higher resolutions, which can strain storage capacities and slow down data processing, particularly when handling large datasets in surveying projects. Additionally, when too many pixels are crammed onto a small sensor, noise can increase, especially in low light conditions, potentially degrading image quality. Balancing pixel count with sensor size is crucial to avoid these issues. Moreover, higher resolution images require more processing power and can consume more battery life, which might limit the drone's flight time.

CMOS sensors play a pivotal role in drone cameras, converting photons into electrical signals to produce images. These sensors are chosen for their efficiency, sensitivity, and speed, which are essential for the dynamic environments drones operate in. The size of the CMOS sensor is critical as it determines the amount of light that can be captured, directly impacting image quality. Larger sensors with larger photodiodes can gather more light, leading to images with less noise and better clarity, particularly important in low-light conditions for accurate mapping.

Pixel size also significantly affects image quality. Larger pixels can capture more light, improving the dynamic range and reducing noise, which is vital for drone photography where lighting conditions can vary widely. In professional drone mapping, 1-inch CMOS sensors are often considered the minimum standard due to their balance of performance and cost. However, for even more demanding tasks, larger sensors like APS-C or full-frame are preferred, offering superior image quality with greater detail retention even under challenging lighting scenarios.

The use of smaller sensors, particularly those below 1 inch, can limit professional mapping capabilities due to their performance in terms of noise and dynamic range. Technologies like Quad Bayer, used in some consumer drones, can give an illusion of higher resolution but do not truly benefit mapping due to their lower effective resolution. This makes most consumer drones less suitable for professional mapping, where precision, detailed settings, and specific calibrations are required.

BRINGING IT ALL INTO FOCUS, THE LENSES:

If you're reading this book because you intend to make photogrammetry or mapping a career choice, then August 1st 1955 is a more significant date than you may realize. Early on that blisteringly hot summer day, a unique event took place. At a remote airbase in the Southern Nevada desert, an aircraft quite unlike any seen before, rolled out of a hanger and taxied to the end of a runway preparing for its first flight. In an era when speed and maneuverability were paramount in the design of aircraft, this uniquely designed airframe stood alone in its capabilities. As the lone pilot lined up at the end of runway 11 and advanced the throttles to full power. The big lumbering bird slowly accelerated down the 23,600-foot runway, and at just over 100 knots, the aircraft struggled into the air for the first time. Its unique design, unstable at low speed and low altitude due to its small thin wing, its small empennage, and its jet engine's high-altitude design characteristics, made flying the unique bird a challenge at low altitude, and the pilot struggled to maintain control of the aircraft throughout the test flight. This, at the time, little-known event, in a section of Edwards Air Force Base called "The Nevada Test and Training Range", a place you more likely know of as "Area 51", would change the world in very profound ways. It was a quantum leap forward in aerospace design, in geopolitics, and most importantly in the context here, in modern aerial photography. The U2 Dragon Lady, a name her pilots would later give her because she was so difficult to fly, ushered in a revolution in aerial photogrammetry. Designed specifically for high altitude photo reconnaissance, she was capable of flying 7000 miles at 75,000 feet, staying airborne for up 15 hours while cruising at 475kts. With the ability to depart from US Air Bases around the globe, overfly anywhere in the Soviet Union and return safely to American soil, she was a revolution in intelligence gathering. She possessed capabilities the Soviets were never able to match.

CIA U2C 1960
Photo: USAF

And while all that aviation technology sounds incredibly impressive, what was even more impressive was the technology in her belly. She possessed a revolution in lens and camera technology. Without the ability to accurately record the data of the areas she overflew she was literally just another airplane "poking the bear". While much faster, and seemingly more capable, reconnaissance aircraft such as the SR71 Blackbird were built to replace her, those impressive aircraft have long been relegated to the history books while the Dragon Lady continues to fly to this day.

The initial camera system installed on the U-2 was the Hycon B camera, designed for high-altitude photographic reconnaissance. This camera allowed the U-2 to take detailed photographs from altitudes above 70,000 feet at 475kts, which was well out of reach of interceptors and anti-aircraft systems of the time. In 1960, another technological leap forward occurred, with the introduction of the Optical Bar Camera.

The Optical Bar Camera (OBC) used in high-altitude reconnaissance missions, such as those carried out by the U-2 and SR-71 Blackbird, is renowned for its sophisticated optical system centered around its powerful lenses. These cameras were integral to the success of Cold War-era surveillance operations due to their ability to capture detailed, wide-area imagery from altitudes exceeding 70,000 feet. The resolution was so high, it is said that, American intelligence officers could determine the type of military base they were photographing by seeing the color of the shoulder boards on the base guards' uniforms.

Key Features of the Lenses in the Optical Bar Camera:

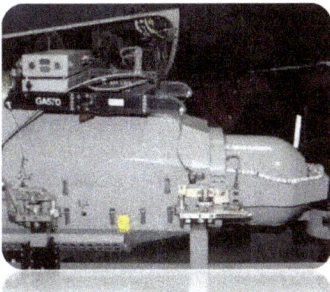

Modern Optical Bar Camera being removed from a U-2S after a reconnaissance flight

High Resolution: The lenses were designed to achieve exceptionally high resolution, capturing fine details over large swaths of terrain. This capability was crucial for identifying small objects and subtle changes in the landscape or military installations from high altitudes.

Wide-Angle Coverage: The OBC utilized wide-angle lenses that allowed for the coverage of extensive areas in a single pass. This wide field of view was essential for strategic reconnaissance, providing a comprehensive overview of large geographic areas.

Specialized Optical Design: The lenses were tailored to minimize optical distortions and aberrations, which are common challenges in high-altitude photography. This design ensured that the images remained sharp and clear across the entire field of view, despite the high speeds and vibrations experienced during flight.

Lightweight Construction: Given the operational altitude and speed of aircraft like the U-2 and SR-71, the lenses were constructed using materials that provided a balance between durability and weight. This lightweight construction was critical to maintain the overall performance of the aircraft.

Adaptability to Film and Conditions: The lenses were compatible with various film types, each suited for different lighting conditions and surveillance needs. This versatility allowed the camera to operate effectively during both day and night missions, adapting to the ambient light available.

Overall, the lenses of the Optical Bar Camera represent a pinnacle of optical engineering, tailored specifically for the demanding conditions of strategic aerial reconnaissance. Their ability to provide clear, detailed images from the edge of space directly supported numerous intelligence and military operations throughout the Cold War.

What, you may ask, does this have to do with me and drones? Well, the simple answer is virtually all drone camera Lens technology has been in some way affected by the design of the Optical Bar Camera lens. From the ability to create undistorted wide images, to the implementation of weight saving measures while maintaining glass quality, to the use of high-speed mechanical shutters, the revolution in photography that began in the late 1950s continues to this day. While the film medium is completely different, the way that film is presented to the sensor has not changed much in 75 years.

The Role of Lenses in Drone Photogrammetry

In drone photogrammetry, lenses are pivotal to the quality and accuracy of captured data, shaping the clarity and precision of the resulting maps and models. Beyond mere optics, they determine how effectively a drone translates aerial perspectives into actionable geospatial insights. This section explores how lens characteristics influence photogrammetric outcomes, offering clear guidance for selecting and optimizing them for diverse missions.

Resolution and Image Quality

High-resolution lenses are the backbone of detailed photogrammetry. These lenses capture images with fine pixel clarity—think 20-45 megapixels on a drone like the DJI Mavic 3 Enterprise—enabling the software to resolve intricate features like tree branches or pavement cracks. The sharper the image, the more precise the 3D triangulation and texture mapping, critical for applications from urban planning to archaeological documentation. A lens with superior resolving power ensures that even at higher altitudes (e.g., 200 feet), the Ground Sample Distance (GSD) remains tight—say, 0.8 cm/pixel—delivering data-rich outputs without sacrificing coverage.

Field of View

The lens's field of view (FOV) dictates how much terrain a single image encompasses, balancing efficiency and detail. Wide-angle lenses (e.g., 20mm focal length) sweep broader areas—ideal for expansive mapping missions like agricultural surveys—capturing hectares in fewer passes. Narrower lenses (e.g., 50mm), with a tighter FOV, focus on smaller, high-detail zones, such as a construction site inspection, where precision trumps breadth. Mission requirements drive this choice: wide-angle for scale, narrow for scrutiny, each optimized to match the project's scope and resolution needs.

Lens Distortion

Distortion can skew photogrammetric accuracy, warping measurements if uncorrected. High-quality lenses minimize issues like chromatic aberration (color fringing) and barrel distortion (curved edges), preserving geometric fidelity—vital when a 1 cm error could shift a boundary line. Modern drones often pair with software (e.g., Pix4Dmapper) to correct residual distortions, adjusting for lens imperfections via calibration profiles. However, software has limits; a premium lens with low distortion—say, a fixed aperture f/2.8—reduces reliance on post-processing, ensuring cleaner raw data for reliable outputs.

Light Adaptation

Lenses must adapt to varying light conditions, from bright midday sun to dim twilight. A lens with a wide aperture (e.g., f/1.8) excels in low light, minimizing noise and maintaining clarity during dawn or dusk flights—key for time-sensitive ecological monitoring. High dynamic range capabilities also help, balancing shadows and highlights in mixed lighting, like a forested site at noon. This versatility ensures consistent image quality across a mission's duration, adapting to environmental shifts without compromising data integrity.

Material and Build Quality

Drone lenses require a delicate balance of durability and lightness. Crafted from robust glass and lightweight alloys, they withstand vibrations and minor impacts—think a Mavic landing on uneven ground—while keeping payload weight low to preserve flight time (e.g., 45 minutes on a Matrice 350). Quality coatings, like anti-reflective layers, reduce flare and enhance contrast, critical under harsh sunlight. This construction ensures reliability across missions, from dusty quarries to humid wetlands, without taxing the drone's performance.

Together, these lens attributes—resolution, FOV, distortion control, light adaptability, and build—form the foundation of effective photogrammetry, turning aerial imagery into precise, practical deliverables.

Focus and Depth of Field

Fixed vs. Adjustable Focus Lenses in Drone Mapping
In drone photogrammetry and mapping, the choice between fixed-focus and adjustable-focus lenses can significantly impact the quality and applicability of the collected data.

Fixed-Focus Lenses: These lenses are set to focus at a specific distance, usually at infinity, ensuring that all objects beyond a certain range are captured in sharp detail. This characteristic makes fixed-focus lenses particularly well-suited for broad area surveys where the primary requirement is consistent, large-scale image capture. With a fixed focus, the drone operator does not need to adjust the focus during flight, which simplifies operations and reduces the risk of capturing out-of-focus images. These lenses are ideal for projects such as agricultural field analysis, forest coverage mapping, and large-scale topographic surveys where the detail requirement is uniform and extensive ground area needs to be covered efficiently.

 Adjustable-Focus Lenses: Offering the ability to adjust the focus distance dynamically, adjustable-focus lenses cater to projects requiring high detail over smaller, specific areas. This flexibility is crucial for applications such as detailed infrastructure inspections, archaeological site documentation, and urban planning, where varying distances from targets can necessitate adjustments to focus for optimal clarity and detail. Adjustable-focus lenses allow the operator to adapt to different subjects at different ranges during a single flight, capturing intricacies that fixed-focus lenses might miss. However, this type of lens requires more skill and interaction from the operator to ensure that images are consistently sharp, which can introduce complexity and potential for human error in focus adjustment during flight.

Choosing the right type of lens depends on the specific needs of the project, with each type offering distinct advantages. Fixed-focus lenses excel in ease of use and consistency over large areas, making them a staple for general mapping. In contrast, adjustable-focus lenses provide the precision needed for close-up, detailed work, essential in inspection and conservation tasks where every detail counts.

By understanding and choosing the right lenses, drone photogrammetry can achieve high levels of precision and effectiveness, just as strategic lens choices enhanced the capabilities of reconnaissance missions in the past. This selection process is crucial for ensuring that the photogrammetric data gathered is reliable, accurate, and suitable for analytical purposes in various fields such as agriculture, construction, and environmental monitoring.

Understanding Ground Sampling Distance

Ground Sampling Distance (GSD) is a critical metric in the realm of drone mapping and photogrammetry, defining the actual distance each pixel, relative to every other pixel in an image, represents. The precision of GSD directly influences the clarity and detail of aerial images, which are vital for accurate measurements and effective data analysis in various applications, from agricultural monitoring to construction site management. By carefully adjusting factors such as the drone's altitude, camera focal length, and the sensor's characteristics, professionals can optimize GSD to suit specific project requirements, ensuring that the resultant maps are both highly detailed and reliable.

Let's explore how GSD is calculated to establish a basic understanding of the math, and art, involved in its calculation. First off, its important to note that, each GSD calculation should be done twice, using each of the 2 formulas provided, and that the lower resolution (higher actual number) of the 2 results should be considered the correct GSD. While that may seem confusing, it seems the common thinking is, it is better to error on the side of caution .

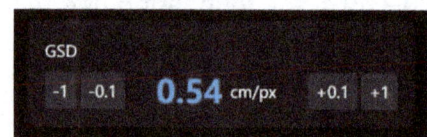

Lets use the Anzu Raptor[4] as an example. It is equipped with a censor that is 4/3in or about 1.33in in diagonal size which has a stated equivalent focal length of 24mm. So, if we need to calculate the GSD of a mission, we would apply the formula of (Altitude (in centimeters) * Sensor Height (in centimeters)) / (focal length (in centimeters) * image Height (in pixels)). The formula being GSD=(FH*SH)/(FL*IH). However, because this calculation is as much art as science, it's necessary to also calculate the alternative formula of Altitude (in centimeters) * Sensor Width (in centimeters) / focal length (in centimeters) * image Width (in pixels). Once both calculations are determined, the most accurate number will be the one exhibiting the lowest resolution or highest actual number. Fortunately, depending on your equipment, once you determine the more accurate formula, you can safely use it and reduce the math. To ease your mind even further, virtually all mapping software calculates this automatically. There is one additional issue to consider, if the drones' specifications are taken at face value you will quickly find that the math doesn't quite work. That's because there's a caveat when it comes to drone technology, in most instances camera specs are skewed to demonstrate ability, not give actual facts about the design. While the Anzu Raptor, and by extension, the Mavic 3 Enterprise, list a "Equivalent focal length" of 24mm for the lens, when we delve a little deeper, we learn that the actual focal length of this particular camera is 12.3mm. so lets examine the chart below to arrive at the necessary data for the calculation.

Focal Length: 12.3mm or 1.23cm

Sensor Width 17.3mm or 1.73cm

---

[4] While the Anzu Raptor is used as an example here, it is important to note that it uses the exact same camera as the DJI Mavic 3 Enterprise, and the same specifications are published on that drone. This is a common industry practice and not indicative of any one manufacturer.

Sensor Height 13mm or 1.3cm

Image Height 3956 Pixels

Image Width 5280 pixels

Assuming our chosen altitude is 61 meters (approximately 200ft), and all of our numbers have been converted to CM, lets do the math.

GSD = $\frac{FlH*SH}{FL*IH}$ in our context: GSD= $\frac{6100*1.3=7930}{1.23*3956=4865.88}$ =7930/4865.88 = 1.629 GSD

But wait, there's more. Because GSD calculation is part art, let's examine the other possible formula.

Again, assuming our chosen altitude is 61 meters (approximately 200ft), and all of our numbers have been converted to CM, lets do the math once more.

GSD = $\frac{FlH*SW}{FL*IW}$ in our context: GSD= $\frac{6100*1.733=10571}{1.23*5280=6494.4}$ =10571/6494.4 = 1.628 GSD

As you can see, both GSD calculations are virtually identical with a very small difference. Since standard practice is to round to the 100th decimal place, either result would give us a GSD of 1.63. As stated earlier in the section, the good news is, virtually all mapping software now calculates GSD as part of the setup process so you will likely never need to use these formulas. However, understanding how they work will better equip you for the missions ahead.

Dewarping - why not to use it.

When creating orthographic maps with drones like the DJI Mavic 3 Enterprise or the DJI Zenmuse P1, vignetting, darkening at the corners of images, can occur due to the optical properties of wide-angle lenses commonly used in mapping cameras. DJI offers a "Dewarping" feature in their systems, which corrects lens distortions, including vignetting, in real-time as images are captured.

However, for ortho missions, DJI recommends leaving this setting turned off. Here's why: Orthographic mapping relies heavily on photogrammetry software (like DJI Terra, Pix4D, or Metashape) to stitch hundreds or thousands of images into a seamless, georeferenced map. These software tools are designed to handle raw, uncorrected images and use sophisticated algorithms to account for lens distortions, including vignetting and the fisheye effect. The key to this process lies in the camera's lens distortion parameters, which are factory-calibrated by DJI for devices like the Mavic 3 Enterprise and Zenmuse P1. These parameters—such as focal length, optical center, and radial and tangential distortion coefficients— are embedded in the image metadata (specifically in the "DewarpData" field under XMP data) when Dewarping is turned off.

When Dewarping is enabled, the camera applies an onboard correction to the images before saving them, effectively flattening the distortion and removing vignetting in real-time. While this might make the images look better visually, it can reduce the accuracy of the photogrammetry process. The onboard correction is less precise than the post-processing correction applied by specialized software, which uses the exact lens parameters to reverse-engineer the distortion with greater fidelity. If the images are pre-corrected with Dewarping, the software may lack the raw data it needs to perform its own, more accurate distortion correction, potentially leading to misalignment, artifacts, or reduced geometric precision in the final orthomosaic.

Additionally, leaving Dewarping off preserves the full field of view captured by the lens, including the slightly distorted edges where vignetting occurs. These edges provide valuable overlap between images, which photogrammetry software uses to improve stitching accuracy. Correcting vignetting onboard might crop or alter these areas, reducing the usable data and potentially compromising the map's quality.

DJI recommends disabling Dewarping for ortho missions because photogrammetry software performs a more accurate and tailored correction using the raw, uncorrected images and their embedded lens parameters. This approach ensures higher precision in the final orthographic map, which is critical for applications like surveying, construction, and geospatial analysis.
Sent from my iPhone

Practical Considerations in Drone Mapping

The importance of matching pixel count with the appropriate sensor size cannot be overstated, especially in applications like drone photogrammetry and mapping, where detail and image quality are crucial. Understanding the trade-offs between resolution and sensor size helps professionals in photogrammetric sciences make informed decisions about the equipment that best meets their specific needs. As drone technology continues to evolve, so too does the capability of onboard cameras, which are increasingly able to capture high-quality data in a wide range of environmental conditions.

In wrapping up our examination of photographic cameras for drones, we have explored the reasons why these tools are essential for precise and effective photogrammetry. The technical aspects—from understanding the exposure triangle to a detailed examination of the nuances of shutter types—form the bedrock upon which successful aerial imaging is built. The quality of images captured by these cameras, influenced by factors such as sensor size, pixel count, and shutter mechanics, directly impacts the accuracy and detail of photogrammetric outputs.

As we transition from RGB photographic cameras, it's important to recognize that the scope of drone capabilities extends beyond visible light photography. The next chapters will explore how drones can be equipped with specialized sensors that open up a new realm of possibilities for data collection and analysis. These include thermal sensors, Multispectral and Hyperspectral sensors, and LiDAR, which allow for data acquisition in environments and conditions where traditional cameras might not be effective.

Thermal sensors, for instance, are invaluable in applications such as search and rescue, wildlife monitoring, and building inspections. They detect heat variations and can be used to assess energy efficiency, locate heat leaks, or identify living beings in obscured conditions. The technical consideration for thermal imaging involves understanding thermal sensitivity, resolution, and the appropriate spectral range for different applications.

LiDAR technology, on the other hand, uses laser light to map physical environments with high precision. This sensor technology is critical for creating detailed 3D models of terrain, measuring forest canopy heights, assisting in flood modeling, and urban planning, among other uses. Technical discussions on LiDAR will focus on pulse rate, scan angle, and how these factors affect the resolution and accuracy of the 3D point cloud data.

By building upon the foundational knowledge of drone cameras discussed in this chapter, we can better appreciate the complexities and capabilities of these advanced sensing technologies. As we move forward, we will explore how integrating these specialized sensors into drone systems transforms the landscape of data collection, providing richer, more varied data that can be leveraged across a multitude of disciplines and industries.

Chapter 4: Specialized Sensors

Where Photographic Cameras Meet Their Limits

Beyond the abilities of standard photographic equipment, specialized sensors play a critical role in extending the capabilities of drone mapping to meet various industry needs. These include multispectral, hyperspectral, LiDAR, and thermal sensors, each serving unique purposes. It's important to note here that, while these sensors have different abilities than RGB cameras, in virtually every instance, an RGB camera is part of the sensor package and deployed in conjunction with these other systems.

This Chapter Will Cover:

- **Multispectral Sensors**: These sensors capture data from across the electromagnetic spectrum, much of which is invisible to the naked eye. They are used in precision agriculture to assess plant health and in environmental monitoring.

- **LiDAR Sensors**: Light Detection and Ranging sensors use laser light to create highly accurate 3D models of the surveyed area. LiDAR is particularly effective in penetrating vegetative cover to map the ground surface underneath, which is crucial for forestry, land management, and topographic survey data in areas of dense vegetation.

- **Thermal Sensors**: Used for environmental monitoring and inspection tasks, these sensors can detect heat variations and energy efficiency in buildings, solar farms, windmills, and power distribution networks. They are also commonly used in search and rescue (SAR), making them highly versatile.

- **Hyperspectral Sensors**: The primary function of hyperspectral sensors is to collect data from hundreds of contiguous wavelengths for each pixel in an image. This enables detailed analysis of the environment because different objects and materials reflect and absorb light differently across various wavelengths. In addition to expanded vegetation health monitoring, hyperspectral sensors are useful for Environmental, Mining, Forestry, Coastal Monitoring, and gas leak detection.

Introduction to Multispectral Imaging

Multispectral cameras represent a significant advancement in remote sensing technology, particularly when integrated with drones. These cameras capture light from several specific, discrete spectral bands across the electromagnetic spectrum, including those beyond the visible light range. This capability allows for detailed analysis of various surface properties and conditions, which is impossible with standard RGB cameras. Multispectral imaging is increasingly used in industries such as agriculture, environmental monitoring, forestry, and even archaeology.

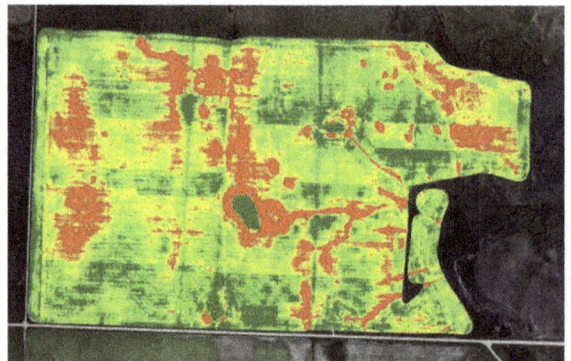

Multispectral Imaging

Basic Principle

Multispectral cameras do not merely capture images; they record data from various wavelengths of light. Each band of light can reveal different features or conditions of the photographed materials or surfaces. For instance, near-infrared light can indicate vegetation health, while thermal infrared can show heat variations in a landscape.

Components of Multispectral Cameras

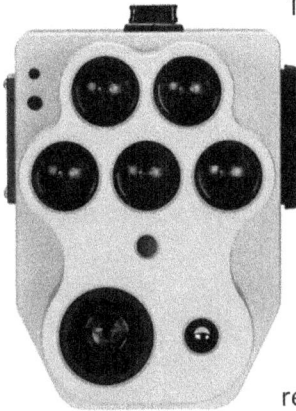

These cameras typically consist of an array of sensors, each tailored to a specific wavelength range. They might also include optical filters that selectively pass light from desired spectral bands. The sophistication of the camera systems can vary, with higher-end models providing greater spectral resolution and more sensitive detection capabilities.

An example of a higher-end model is the MicaSense Altum-PT. Here is a quote from the MicaSense website about the camera:

"The Altum-PT is an optimized 3-in-1 solution for advanced remote sensing and agricultural research. This solution seamlessly integrates an ultra-high-resolution, 12MP panchromatic imager, a built-in 320x256 radiometric thermal imager, and five discrete spectral bands to obtain synchronized outputs such as RGB color, crop vigor, heat maps, and high-resolution panchromatic in just one flight. Higher resolution also means more analytical capabilities and applications: discern issues at the plant level, even in the early growth stages, including early-stage stand counting. Altum-PT also features a global shutter for distortion-free results on any platform, open APIs and standard data formats for compatibility with multiple processing platforms, and a new professional removable storage standard in CFexpress for two captures per second and the fastest turnaround time between flights."

What Each Sensor Does

Here's a description of each camera or spectral band in the MicaSense Altum-PT multispectral sensor and its use in multispectral agricultural analysis, based on the available information about the sensor's capabilities for precision agriculture, crop health monitoring, and environmental assessment:

- **Green (560 nm center, 27 nm bandwidth):** The green band captures light in the visible green spectrum, which is essential for assessing overall plant health and vigor. In multispectral agricultural analysis, it helps identify chlorophyll content and plant density because healthy vegetation reflects green light strongly. This band is often used in combination with other bands to calculate vegetation indices like the Normalized Difference Vegetation Index (NDVI) to monitor crop growth, detect early signs of stress, and evaluate canopy cover, aiding farmers in optimizing irrigation, fertilization, and pest management.

- **Blue (475 nm center, 32 nm bandwidth):** The blue band captures light in the visible blue spectrum, which is useful for detecting plant stress, nutrient deficiencies, and water imbalances. In agricultural analysis, it helps identify early signs of disease or damage that might not be visible in other bands because stressed or unhealthy plants often reflect less blue light. This band is also used in vegetation indices and for distinguishing between soil and vegetation, supporting precise mapping of crop health and guiding targeted interventions like variable rate applications of inputs.

- **Red (668 nm center, 14 nm bandwidth)**: The red band captures light in the visible red spectrum, which is critical for assessing plant health and photosynthetic activity. In multispectral analysis, it's used to detect chlorophyll absorption because healthy plants absorb red light for photosynthesis. This band is a key component of vegetation indices like NDVI, helping farmers monitor crop vigor, identify nutrient deficiencies, and detect diseases or pests early. It's particularly useful for distinguishing between healthy and stressed vegetation, aiding in irrigation and yield optimization decisions.

- **Red Edge (717 nm center, 12 nm bandwidth)**: The red edge band captures light at the transition zone between red and near-infrared, making it highly sensitive to changes in plant health and chlorophyll content. In agricultural analysis, it's used to detect subtle variations in vegetation health, such as early stress, nutrient deficiencies, or disease, before they become visible in other bands. This band is especially valuable for precision agriculture because it improves the accuracy of crop and weed mapping (e.g., distinguishing winter wheat from weeds like chickweed and hairy buttercup) and supports advanced vegetation indices like the Normalized Difference Red Edge Index (NDRE) for detailed plant physiology analysis and yield prediction.

- **Near Infrared (NIR) (842 nm center, 57 nm bandwidth)**: The near-infrared band captures light beyond the visible spectrum, which healthy vegetation reflects strongly due to its cellular structure. In multispectral agricultural analysis, NIR is crucial for assessing plant health, biomass, and water content. It's a primary component of vegetation indices like NDVI, enabling farmers to monitor crop vigor, detect stress, and evaluate irrigation efficiency. NIR data helps identify areas of the field with poor growth or potential disease, supporting targeted management practices to improve yield and resource use.

- **Thermal (LWIR) (7.5–13.5 μm, radiometrically calibrated)**: The thermal long-wave infrared (LWIR) camera measures surface temperatures, providing insights into crop canopy and soil temperature variations. In agricultural analysis, it's used to detect water stress, irrigation inefficiencies, and plant physiological status by identifying temperature differences that indicate moisture levels or stress (e.g., cooler areas with excessive irrigation or warmer areas with insufficient water). Thermal imaging also helps identify irrigation system issues like clogs or leaks and supports early detection of disease or pest infestations through temperature anomalies, aiding in precise water management and crop health monitoring.

- **Panchromatic (12.4 MP high-resolution sensor)**: The panchromatic band captures high-resolution imagery across a broad spectrum of visible light, typically used to enhance the spatial resolution of multispectral data through a process called pan-sharpening. In agricultural analysis, it provides detailed, high-resolution RGB and multispectral outputs, improving the accuracy of plant-level applications like early-stage crop counting, weed identification, and advanced vegetation research. Its global shutter ensures distortion-free images, making it ideal for machine learning tasks and precise mapping of crop health, yield estimation, and field variability at leaf-level resolutions.

Applications of Multispectral Cameras in Drones

Agriculture

- **Crop Health Monitoring**: Multispectral imaging can detect plant stress before it becomes apparent visually. By measuring the light absorbed and reflected by crops, these cameras can identify areas of poor health due to pests, diseases, insufficient nutrients, or localized drought.

- **Water Usage and Irrigation Planning**: Analysis of multispectral images helps determine water stress in crops, enabling more efficient water management and irrigation planning.

- **Yield Optimization**: Farmers use data from multispectral sensors to assess the vigor of their crops, which helps predict yield and optimize inputs like fertilizers and pesticides based on the specific needs of different crop zones.

Environmental Conservation

- **Habitat Analysis**: Multispectral cameras on drones can monitor vegetation types and health across large and inaccessible areas, aiding in the management of forests, wetlands, and other natural habitats.

- **Water Quality Assessment**: Multispectral cameras can help assess water quality parameters such as turbidity, chlorophyll levels, and the presence of pollutants by capturing data on different bands, including the visible and near-infrared.

- **Disaster Management and Response**: These cameras are instrumental in assessing damage from disasters like floods or wildfires, providing critical information for response and recovery efforts.

Forestry

- **Forest Health Monitoring**: Like agriculture, multispectral imaging can assess forest health and detect disease outbreaks, pest infestations, or drought stress.

- **Species Identification**: Different species of trees can be identified based on their spectral signatures, aiding in biodiversity assessments.

- **Timber Volume Estimation**: Multispectral data, combined with 3D mapping techniques, can be used to estimate timber volume and plan harvests more effectively.

Archaeology

- **Site Discovery**: Multispectral imaging can help archaeologists locate buried features by detecting anomalies in vegetation and soil properties that may indicate man-made structures or disturbances below the surface.

- **Documentation of Sites**: Drones equipped with multispectral cameras provide a means to document the current condition of archaeological sites over broad areas, aiding in preservation efforts.

Technological Considerations

Integration with Drone Technology

The lightweight and compact design of modern multispectral cameras makes them ideal for drone integration. Drones can cover large areas quickly and provide vantage points that are not feasible with ground-based observations.

Data Processing and Analysis

The data captured by multispectral cameras are complex and require specialized software for processing and analysis. This software typically includes tools for calibrating the data to correct for atmospheric interference and for combining images from different spectral bands into usable formats.

AI and Machine Learning

Advanced algorithms and machine learning techniques are increasingly being used to interpret multispectral data automatically. These tools can classify surface types, detect anomalies, and even predict trends based on historical data.

Future Trends

As drone and sensor technology continues to evolve, the use and capabilities of multispectral cameras are expected to expand. Improvements in sensor design might lead to even finer spectral resolutions, while advances in drone technology could allow for longer flight times and more stable platforms, enhancing the quality and scope of multispectral imaging.

Use Case: Corn Farming in Indiana

Fully equipped with gimbal and software, this particular sensor will cost north of $20,000 (May 2024 pricing), which does not include the drone. Among the sUASs these cameras are compatible with are DJI Matrice M350 RTK drones, which can add another $15,000 to $20,000 to the end price. While that price may seem high to some, when employed correctly with a spray drone such as a DJI Agras series, these technologies can increase crop yields by as much as twenty percent (Source: Agremo Field Analytics).

In real terms, consider that a farm in central Indiana produces an average of 253 bushels per acre of corn. Increasing the yield by just ten percent—half of the stated potential—increases the yield to 278 bushels. At $4.83 per bushel (market price May 2024), that translates to an additional $122.20 per acre of increased revenue. On a 1000-acre farm, that represents a revenue increase of $122,200 annually. Now, consider that the M350 can fly more than 2000 acres per day. If a farm is flown once a week for the 78-day growing season, that's 11 flights. If half of those flights detect an issue, the Agras drone can then be deployed to treat the problem. If the farmer pays $1 per acre—or $1000—for a monitoring flight one day per week and an average of $2000 per corrective flight for the Agras drone, the total cost to increase his yield is $22,000 per season. That represents a profit of $100,200 per year. This is why seed companies have been deploying multispectral drones for almost a decade. Moreover, the practice is growing exponentially.

If the Agras spray drone has piqued your interest, be aware that it requires additional licensing to operate and is typically operated directly by seed and pest control companies.

Multispectral Camera Drones in Use Today

Here is an overview of some prominent drone-based multispectral cameras, highlighting their specifications and features, which are critical for tasks like agricultural monitoring, environmental management, and land surveying:

DJI Mavic 3 Multispectral (Mavic 3M)

- **Camera Specs**: The camera features four 5MP multispectral cameras covering the green, red, red edge, and near-infrared wavelengths. It also includes a high-resolution 20MP RGB camera with a 4/3 CMOS sensor.

- **Applications**: This camera array is ideal for precision agriculture, enabling detailed crop health monitoring and management. It facilitates the generation of NDVI maps and other vegetation indices, which are crucial for assessing plant health.

- **Additional Features**: The drone includes an RTK module for centimeter-level positioning accuracy and can cover up to 2 square kilometers in a single flight thanks to its 43-minute battery life. It's designed to connect seamlessly with the DJI Terra or DJI SmartFarm platform for detailed data analysis.

AgEagle eBee X

- **Camera Specs**: This setup is equipped with the Duet M multispectral camera, which captures high-resolution RGB and multispectral images. It is tailored for comprehensive field analysis and crop monitoring.

  - **Applications**: Primarily used in precision agriculture for crop mapping and monitoring, it assists in decision-making regarding irrigation, fertilization, and pest management.

- **Flight Performance**: Known for its lightweight, durable design, it can autonomously navigate predefined flight paths with up to 55 minutes of flight time on a single battery, enhancing its utility for extensive agricultural data collection.

DJI Phantom 4 Multispectral

- **Camera Specs**: Integrates a 6-camera RGB and multispectral imaging system that includes an RGB camera and five 2MP cameras covering the blue, green, red, red edge, and near-infrared bands.

- **Applications**: Suited for smart agriculture and land management, this drone can generate precise data for crop health analysis and environmental monitoring.

- **Flight Features**: It offers a 27-minute flight time and can collect detailed data for effective vegetation management and crop treatment.

MicaSense Altum-PT

The MicaSense Altum-PT is somewhat drone-agnostic. There are multiple configurations available for drones such as the DJI Matrice 300/350, the AgEagle eBee, the Vision Aerial Vector, the Inspired Flight series of drones, the FreeFly Astro, the WISPR Ranger Pro, and the Arcsky X55, just to name a few.

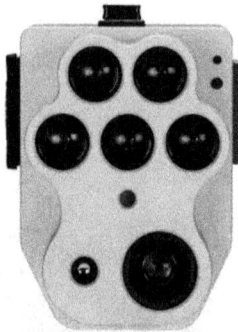

- **Camera Specs**: The Altum-PT integrates a thermal, multispectral, and high-resolution panchromatic imager into one compact package. It captures five discrete spectral bands (blue, green, red, red edge, and near-infrared) and provides thermal imaging capabilities, all processed by an onboard processor.

- **Applications**: This camera is designed for advanced remote sensing and agricultural research. It offers detailed insights into plant health monitoring, irrigation management, and disease detection. Its thermal sensor is pivotal for water stress analysis.

- **Additional Features**: With integrated GPS and advanced radiometric calibration, the Altum-PT ensures that each image is precisely geotagged and calibrated for accurate measurements. The camera is also capable of synchronized captures, meaning that the multispectral and thermal images align perfectly, allowing for complex analysis.

- **Compatibility**: The Altum-PT is compatible with a variety of drones and is particularly well-suited for use in fixed-wing UAVs for large-scale operations due to its lightweight and low power consumption. It provides essential data for precision agriculture, environmental monitoring, and forestry management.

The MicaSense Altum-PT stands out for its ability to capture high-resolution multispectral and thermal data. It provides a comprehensive imaging solution that enhances decision-making in various fields, particularly precision agriculture. Its robust feature set makes it a powerful tool for comprehensive environmental monitoring and management.

These cameras are designed to support a variety of spectral ranges and are adaptable for different uses. They are particularly noted for their energy efficiency and lightweight design, making them suitable for long flights typical of extensive surveying and monitoring tasks. Each of these models exemplifies how multispectral imaging technology integrated into drones is transforming data collection across various industries, making it more efficient and precise.

When mounted on drones, multispectral cameras open up new possibilities across various fields by providing detailed spectral data that is not visible to the human eye. This technology has the potential to revolutionize precision agriculture, environmental monitoring, forestry management, and many other areas by providing precise, actionable data. As these technologies evolve, they will quickly become more accessible and widely used, offering profound insights and efficiencies in managing natural resources and planning agricultural and environmental initiatives.

LiDAR Technology in Drone Mapping

LiDAR (Light Detection and Ranging) technology is a critical component of modern geospatial data collection, particularly when integrated with drone technology. By providing precise distance measurements through laser light, LiDAR offers a unique set of capabilities that are invaluable in a wide array of applications, from forestry management to urban planning. This section explores the principles of LiDAR, its advantages in drone mapping, and its diverse use cases across various industries.

Principles of LiDAR

LiDAR technology works by emitting laser light towards the target area and measuring the time it takes for the light to return after reflecting off surfaces. This time-of-flight (TOF) principle allows LiDAR systems to calculate distances with high precision. A LiDAR sensor typically includes a laser, a scanner, and a specialized GPS receiver. Airborne LiDAR systems, especially those mounted on drones, combine this data with positional data from the GPS and inertial measurement units (IMUs) to create accurate three-dimensional information about the Earth's surface and its features.

LiDAR is particularly well-suited for topographical data collection because it can produce precise 3D models of the Earth's surface at ground level as well as its overlaying vegetation. Unlike photographic images, which can be obscured by weather conditions or low light, LiDAR can operate in most weather conditions and at various times of the day, making it incredibly reliable for consistent data collection.

Advantages of Using LiDAR in Drone Mapping

LiDAR technology, when used in conjunction with drones, brings several substantial benefits to the field of geospatial data collection:

- **Penetration of Vegetation**: LiDAR's most significant advantage is its ability to penetrate dense vegetation. LiDAR lasers can reach the ground level beneath tree canopies, providing accurate elevation data and ground cover details essential for environmental and forestry management. This capability is critical in areas where other data collection methods might be hindered by foliage.

- **High Accuracy and Resolution**: LiDAR data is extremely precise, offering detailed and accurate measurements that are invaluable in many technical and scientific applications. This precision is crucial for creating reliable models for infrastructure projects, hazard preparedness, and historical site documentation.

- **Rapid Data Collection**: Drones equipped with LiDAR sensors can cover large areas more quickly than ground surveys, significantly reducing the time and cost associated with data collection. This efficiency is particularly beneficial in projects with tight deadlines or in environments that are difficult to access or where ground access may be restricted for environmental reasons.

- **3D Modeling**: LiDAR is excellent for creating detailed 3D models of the surveyed areas. These models are useful in various applications, including simulation environments, virtual reality, and comprehensive urban planning.

Use Cases of LiDAR Applications

The versatility of LiDAR technology allows it to be applied in numerous fields, demonstrating its extensive capabilities and benefits:

Topographical and Boundary Surveying

In the domain of land surveying, LiDAR technology plays a pivotal role in conducting topographical and boundary surveys. These surveys are essential for delineating property lines, planning construction projects, and ensuring legal compliance in land development. LiDAR's capability to capture precise elevation data and generate accurate 3D models of the terrain makes it an invaluable tool in these contexts.

- **Topographical Surveys**: Utilizing LiDAR, surveyors can quickly and accurately map the contours and features of a land area. This detailed topographical data is crucial for architects, engineers, and construction managers in planning and designing buildings, roads, and drainage systems. The high-resolution elevation data helps in identifying natural drainage paths, planning grading, and other land modifications necessary for development projects.

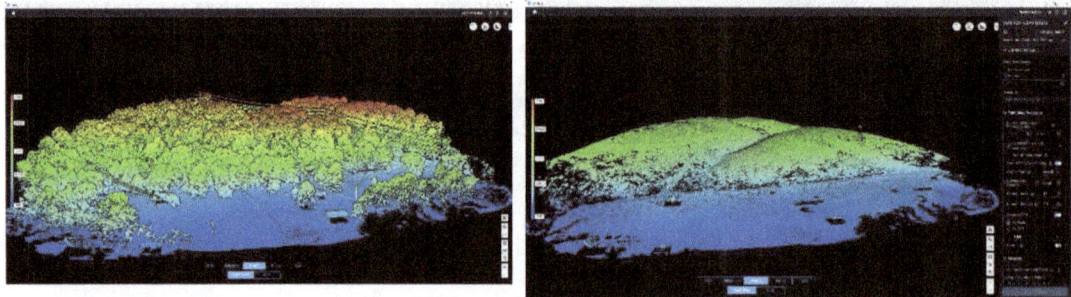

- **Boundary Surveys**: For boundary surveys, LiDAR provides a clear advantage by accurately mapping property boundaries over varied terrain. The technology can penetrate vegetation to reveal the ground surface, ensuring that boundary lines are drawn accurately, even in heavily wooded or overgrown areas. This is crucial for resolving property disputes, updating land records, and facilitating real estate transactions.

By integrating LiDAR data into the surveying process, professionals can achieve a higher level of precision and efficiency, reducing the time and labor traditionally required for land surveying. Moreover, the ability to capture comprehensive data in a single pass minimizes disturbances to the surveyed land, making LiDAR an eco-friendly choice for modern surveying needs.

Forestry Management

In forestry, LiDAR is used to create detailed biomass estimates, assess forest canopy health, and map forest structures for conservation and management purposes. The data collected helps measure forest densities, tree heights, and terrain variations, which are crucial for sustainable forest management.

Flood Modeling and Risk Assessment

LiDAR's ability to produce accurate elevation data is vital in hydrology, especially in creating detailed flood models and watershed analyses. These models help predict water flow patterns and potential flooding areas, assisting in disaster preparedness and mitigation strategies.

Urban Planning

Urban planners utilize LiDAR data to model city landscapes, assess infrastructure projects, and plan new developments. The detailed topographical data supports everything from constructing new buildings to improving transportation systems, ensuring that changes to the urban environment are made with precision and foresight.

LiDAR technology has transformed the landscape of geospatial data collection, bringing unparalleled precision to the mapping, surveying, and analysis of natural and built environments. As drone technologies continue to evolve, the integration of LiDAR systems enhances their utility and expands their application across industries. The future of LiDAR in drone mapping looks promising, with ongoing advancements likely to unlock even more sophisticated capabilities and broader use cases, reinforcing LiDAR's role as a cornerstone technology in modern mapping and surveying.

Use Case: Where LiDAR Pays the Bills

LiDAR sensors are particularly valuable in projects where penetration of dense vegetation is needed or where fine-scale topographic details are required. They are often deployed in geological surveys, forest management to assess biomass and structure, and large-scale infrastructure projects for detailed site analysis. Access to relatively inexpensive LiDAR technology is changing the game in topographical surveys. What once took two surveyors two weeks to

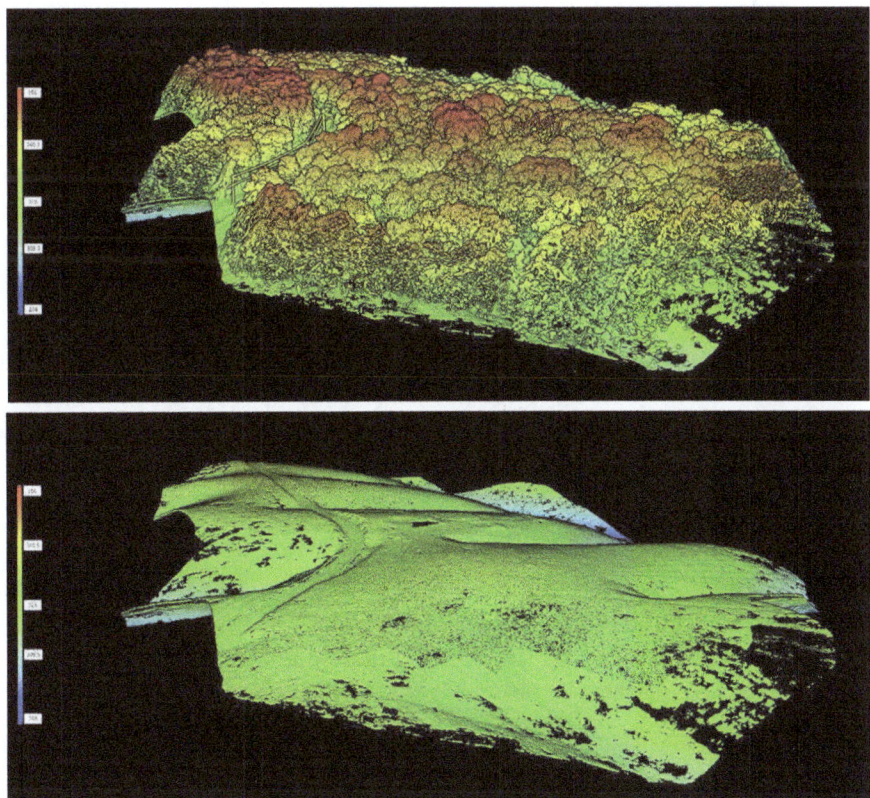

perform at resolutions of 1 point per 525 square feet can now be done by a single trained pilot at a resolution of over 500 points per square foot in less than one day. There are few industries where a $50,000 equipment investment can reduce labor costs by as much as a million dollars per year (one drone pilot can effectively replace up to 20 laborers) while at the same time increasing efficiency. When viewed from this perspective, it becomes very evident why more and more surveyors are entering the world of aerial surveys.

Comprehensive LiDAR Unit Comparison for Drone Mapping

LiDAR (Light Detection and Ranging) units are essential for high-precision drone mapping, providing accurate 3D spatial data for various applications. This section offers a detailed comparison of ten leading LiDAR units suitable for drone-based mapping: Riegl miniVUX-3UAV, Quanergy M8, Ouster OS1, Hesai PandarXT-32, DJI Zenmuse L2, YellowScan Mapper, Leica BLK2FLY, Phoenix LiDAR XT32, Teledyne Optech Maverick, and Rock 3 Pro. Each unit is evaluated based on consistent criteria to facilitate direct comparison, focusing on technical specifications, strengths, weaknesses, and practical considerations. The aim is to provide end users with comprehensive insights to select the optimal LiDAR unit for their drone mapping needs.

Evaluation Criteria

For each LiDAR unit, we assess the following aspects:

- **Number of Lasers (Channels):** Determines the density and coverage of point clouds.
- **Returns:** The number of returns per laser pulse, affecting the ability to penetrate vegetation or capture complex surfaces.
- **Mapping Speed:** The rate of data collection, impacting efficiency for large-scale projects.
- **Proprietary Software:** Dependency on manufacturer-specific software for initial data processing and its implications.
- **Cost vs. Resolution:** The balance between price and the quality/detail of the output data.
- **Unique Strengths and Weaknesses:** Individualized features or limitations that distinguish each unit.

LiDAR Units Compared

1. Riegl miniVUX-3UAV

The Riegl miniVUX-3UAV is a premium LiDAR unit designed for professional-grade drone mapping.

- **Number of Lasers:** Equivalent to 100 channels via a rotating mirror system, delivering exceptionally dense point clouds.
- **Returns:** Up to 5 returns per pulse, excelling in penetrating dense vegetation and capturing ground surfaces beneath canopies.
- **Mapping Speed:** Up to 200,000 points per second, with high accuracy enabling rapid coverage (e.g., 20 hectares in ~20 minutes at 80m altitude).

- **Proprietary Software:** Requires Riegl's RiPROCESS suite for initial data processing, which is powerful but complex and costly. Integration with third-party tools is possible but requires expertise.
- **Cost vs. Resolution:** Priced at $50,000–$70,000, the miniVUX-3UAV offers sub-centimeter resolution (0.1–0.2 cm at 50m), ideal for applications demanding extreme precision, such as archaeological site mapping or utility corridor inspections.
- **Strengths:**
  - High return count enhances performance in challenging environments like forests or urban canyons.
  - Superior accuracy (±1 cm) and precision make it a top choice for survey-grade applications.
  - Compact (1.6 kg) and optimized for integration with professional drones like the Freefly Alta X.
- **Weaknesses:**
  - High cost limits accessibility for smaller operations.
  - RiPROCESS software has a steep learning curve and additional licensing fees.
  - Higher power consumption requires robust drone power systems.

End-User Expectations: The miniVUX-3UAV is tailored for professionals needing survey-grade data, such as civil engineers or environmental scientists. Users must be prepared for significant upfront and ongoing software costs, but the unit's resolution and reliability justify the investment for high-stakes projects.

2. Quanergy M8

The Quanergy M8 is a solid-state LiDAR unit designed for durability and versatility.

- **Number of Lasers:** 8 channels.
- **Returns:** Up to 3 returns per pulse, adequate for moderate vegetation penetration.
- **Mapping Speed:** Up to 420,000 points per second, enabling fast coverage (e.g., 15 hectares or 37 acres in ~12 minutes at 60m altitude).
- **Proprietary Software:** Uses Quanergy's Qortex for initial processing, which is user-friendly but requires a subscription for full functionality. Integration with open-source tools like PCL (Point Cloud Library) is supported.
- **Cost vs. Resolution:** Priced at $12,000–$15,000, the M8 offers a resolution of ~0.3 cm at 50m, suitable for mid-range applications like road corridor mapping.
- **Strengths:**
  - Solid-state design enhances durability, reducing maintenance needs in rugged environments.
  - High point rate supports efficient large-area mapping.
  - Wide operating temperature range (-20°C to 60°C), ideal for extreme conditions.
- **Weaknesses:**
  - Limited channel count restricts point cloud density compared to higher-end units.
  - Qortex subscription costs add to long-term expenses.
  - Heavier (900g) than some competitors, requiring more robust drone platforms.

End-User Expectations: The M8 is well-suited for mid-range mapping tasks, such as transportation infrastructure or mining site surveys. Its durability and speed are key advantages, but users should budget for software subscriptions and ensure their drones can handle the unit's weight.

3. Ouster OS1

The Ouster OS1 is a high-performance LiDAR unit known for its versatility and data quality.

- **Number of Lasers:** 64 channels.
- **Returns:** Up to 4 returns per pulse, excellent for complex environments like forests or urban areas.
- **Mapping Speed:** Up to 1,310,000 points per second, enabling rapid coverage (e.g., 30 hectares in ~15 minutes at 80m altitude).
- **Proprietary Software:** Relies on Ouster's Ouster Studio for initial processing, which is intuitive but limited for advanced workflows. Data is easily exported to tools like FARO Scene or Bentley Pointools.
- **Cost vs. Resolution:** Priced at $20,000–$25,000, the OS1 delivers a resolution of ~0.2 cm at 50m, competitive for high-detail applications like structural inspections.
- **Strengths:**
    - High channel count and return capability produce dense, detailed point clouds.
    - Wide field of view (360° horizontal, 45° vertical) minimizes flight passes.
    - Robust build (IP68 rating) ensures reliability in harsh conditions.
- **Weaknesses:**
    - Higher cost than entry-level units, potentially prohibitive for small operations.
    - Ouster Studio requires additional tools for comprehensive post-processing.
    - Weight (1.2 kg) demands more powerful drones, limiting compatibility with smaller platforms.

End-User Expectations: The OS1 is a strong choice for professionals tackling large-scale, high-detail projects, such as urban planning or forestry management. Its high point rate and resolution are major assets, but users should plan for additional software and drone compatibility considerations.

4. Hesai PandarXT-32

The Hesai PandarXT-32 is a mid-range LiDAR unit optimized for drone and automotive applications.

- **Number of Lasers:** 32 channels.
- **Returns:** Up to 3 returns per pulse, suitable for moderate vegetation penetration.
- **Mapping Speed:** Up to 640,000 points per second, enabling efficient coverage (e.g., 20 hectares in ~15 minutes at 70m altitude).
- **Proprietary Software:** Uses Hesai's PandarView for initial processing, which is free but basic. Integration with third-party tools like TerraSolid or LAStools is common.
- **Cost vs. Resolution:** Priced at $15,000–$18,000, the PandarXT-32 offers a resolution of ~0.25 cm at 50m, balancing cost and quality for applications like pipeline monitoring.
- **Strengths:**
    - Good balance of channel count and point rate for mid-range applications.
    - Compact (1.1 kg) and energy-efficient, compatible with a wide range of drones.
    - High angular resolution (0.09°) enhances detail in point clouds.

- **Weaknesses:**
  - Three-return limit constrains performance in dense environments.
  - PandarView's limited functionality requires additional software for advanced processing.
  - Slightly narrower field of view (360° horizontal, 31° vertical) than some competitors.

End-User Expectations: The PandarXT-32 is a versatile option for mid-range mapping tasks, such as environmental monitoring or infrastructure surveys. Its balance of cost, resolution, and compatibility makes it appealing, but users should anticipate additional software costs for complex workflows.

5. DJI Zenmuse L2

The DJI Zenmuse L2 is an integrated LiDAR system designed for seamless use with DJI drones, offering a turnkey solution for mapping.

- **Number of Lasers:** Equivalent to 6 channels (uses Livox Avia sensor with non-repetitive scanning).
- **Returns:** Up to 5 returns per pulse, excellent for penetrating dense vegetation.
- **Mapping Speed:** Up to 240,000 points per second, suitable for small to medium areas (e.g., 10 hectares in ~12 minutes at 50m altitude).
- **Proprietary Software:** Relies on DJI Terra for initial processing, which is user-friendly and supports one-click post-processing but requires a subscription for full features. Integration with third-party tools is limited.
- **Cost vs. Resolution:** Priced at $13,000–$15,000, the L2 offers a resolution of ~0.4 cm at 50m, competitive for its price but less precise than premium units.
- **Strengths:**
  - Seamless integration with DJI platforms (e.g., Matrice 350 RTK) simplifies setup and operation.
  - Includes a 20 MP RGB camera for colorized point clouds, enhancing visual interpretation.
  - High return count supports applications like forestry and powerline inspections.
- **Weaknesses:**
  - Limited channel count restricts point cloud density compared to high-end units.
  - DJI Terra's subscription model increases long-term costs.
  - Performance is optimized for daytime conditions, potentially limiting use in low-light environments.

End-User Expectations: The Zenmuse L2 is ideal for users seeking an easy-to-use, integrated solution for small to medium-scale projects, such as topographic mapping or infrastructure inspections. Its affordability and DJI ecosystem integration are major advantages, but users should budget for software subscriptions and plan missions for optimal lighting conditions.

6. YellowScan Mapper

The YellowScan Mapper is a lightweight, versatile LiDAR system designed for integration with various UAVs.

- **Number of Lasers:** Equivalent to 6 channels (uses Livox Avia sensor).
- **Returns:** Up to 3 returns per pulse, adequate for moderate vegetation penetration.

- **Mapping Speed:** Up to 240,000 points per second, enabling coverage of small areas (e.g., 5 hectares in ~8 minutes at 50m altitude).
- **Proprietary Software:** Uses YellowScan CloudStation for initial processing, which is intuitive and designed for surveyors but requires a license. Supports export to standard formats for third-party tools like Global Mapper.
- **Cost vs. Resolution:** Priced at $20,000–$25,000, the Mapper delivers a resolution of ~0.5 cm at 50m, suitable for applications like environmental monitoring but less competitive for high-precision tasks.
- **Strengths:**
  - Lightweight (1.1 kg) and compatible with multiple drones, including DJI Matrice 300 and Acecore platforms.
  - IP67 rating ensures durability in rugged environments.
  - High precision (2–5 cm accuracy) for topographic surveys and coastal monitoring.
- **Weaknesses:**
  - Limited channel count and return capability reduce performance in dense environments.
  - CloudStation's licensing costs add to expenses.
  - Shorter range (up to 100m) restricts use for large-scale or high-altitude mapping.

End-User Expectations: The YellowScan Mapper is well-suited for users needing a versatile, weatherproof solution for small-scale mapping tasks, such as mining or coastal surveys. Its ease of integration is a key advantage, but users should plan for licensing costs and consider its range limitations for larger projects.

7. Leica BLK2FLY

The Leica BLK2FLY is an autonomous LiDAR system optimized for detailed architectural and infrastructure modeling.

- **Number of Lasers:** Equivalent to 32 channels (SLAM-based scanning with multi-target resolution).
- **Returns:** Up to 4 returns per pulse, suitable for complex urban environments.
- **Mapping Speed:** Up to 700,000 points per second, enabling efficient coverage (e.g., 10 hectares in ~10 minutes at 50m altitude).
- **Proprietary Software:** Uses Leica's Cyclone REGISTER 360 for initial processing, which is robust but expensive and complex. Integration with third-party tools is supported but requires expertise.
- **Cost vs. Resolution:** Priced at $80,000–$100,000, the BLK2FLY offers a resolution of ~0.2 cm at 50m, ideal for high-precision applications like as-built modeling.
- **Strengths:**
  - Autonomous flight capabilities simplify operation for indoor and outdoor mapping.
  - Full-dome scanning (360° horizontal and vertical) ensures comprehensive data capture.
  - SLAM technology enhances performance in GPS-denied environments, such as indoor spaces.
- **Weaknesses:**
  - High cost limits accessibility for smaller operations.
  - Heavy (2.5 kg), requiring robust drones like the Freefly Alta X.
  - Cyclone software's complexity and cost may deter non-professional users.

End-User Expectations: The BLK2FLY is tailored for professionals in architecture, engineering, and construction needing high-precision 3D models. Its autonomous operation and SLAM capabilities are unique advantages, but users must be prepared for significant costs and robust drone requirements.

8. Phoenix LiDAR XT32

The Phoenix LiDAR XT32 is a high-performance unit designed for advanced topographic and environmental mapping.

- **Number of Lasers:** 32 channels (Hesai XT32 sensor).
- **Returns:** Up to 3 returns per pulse, adequate for moderate vegetation penetration.
- **Mapping Speed:** Up to 640,000 points per second, enabling efficient coverage (e.g., 20 hectares in ~15 minutes at 70m altitude).
- **Proprietary Software:** Uses Phoenix LiDARMill for initial processing, which is automated but requires a subscription (~$10,000/year). Supports export to tools like TerraSolid for advanced workflows.
- **Cost vs. Resolution:** Priced at $30,000–$40,000, the XT32 offers a resolution of ~0.25 cm at 50m, balancing cost and quality for applications like forestry and powerline inspections.
- **Strengths:**
  - Versatile mounting options for drones, vehicles, or backpacks.
  - High point rate and range (up to 400m) support large-scale mapping.
  - Reliable performance when paired with a quality IMU (e.g., Applanix APX-15).
- **Weaknesses:**
  - Three-return limit constrains performance in dense environments.
  - LiDARMill's subscription model increases long-term costs.
  - Weight (1.5 kg) requires robust drone platforms.

End-User Expectations: The XT32 is a strong choice for mid-to-large-scale mapping projects, offering flexibility and high performance. Users should budget for software subscriptions and ensure compatibility with their drone's payload capacity.

9. Teledyne Optech Maverick

The Teledyne Optech Maverick is a portable LiDAR unit designed for mobile mapping in harsh environments.

- **Number of Lasers:** 32 channels.
- **Returns:** Up to 4 returns per pulse, suitable for complex terrains and urban settings.
- **Mapping Speed:** Up to 800,000 points per second, enabling rapid coverage (e.g., 25 hectares in ~15 minutes at 60m altitude).
- **Proprietary Software:** Uses Teledyne's LMS Pro for initial processing, which is robust but requires a license. Integration with third-party tools like Global Mapper is supported.
- **Cost vs. Resolution:** Priced at $40,000–$50,000, the Maverick offers a resolution of ~0.2 cm at 50m, ideal for high-detail applications like mobile mapping.
- **Strengths:**
  - Portable design (1.3 kg) supports multiple platforms, including drones and vehicles.
  - Robust against environmental challenges (IP65 rating), suitable for harsh conditions.

- High point rate enhances efficiency for large-area surveys.
- **Weaknesses:**
    - Higher cost than mid-range units, potentially limiting accessibility.
    - LMS Pro's licensing costs add to expenses.
    - Range (up to 250m) is shorter than some competitors, limiting high-altitude mapping.

End-User Expectations: The Maverick is well-suited for professionals needing a durable, high-performance solution for mobile mapping in challenging environments, such as urban or industrial sites. Users should plan for licensing costs and consider its range limitations for large-scale projects.

10. Rock 3 Pro

The Rock 3 Pro is a high-end LiDAR unit designed for long-range, high-precision drone mapping.

- **Number of Lasers:** 40 channels (Hesai Pandar40 sensor).
- **Returns:** Up to 4 returns per pulse, excellent for complex environments like forests or urban areas.
- **Mapping Speed:** Up to 1,200,000 points per second, enabling rapid coverage (e.g., 35 hectares in ~15 minutes at 80m altitude).
- **Proprietary Software:** Uses Rock Robotic's SpatialExplorer Pro for initial processing, which is automated and user-friendly but subscription-based. Supports export to tools like LiDAR360.
- **Cost vs. Resolution:** Priced at $35,000–$45,000, the Rock 3 Pro offers a resolution of ~0.15 cm at 50m, competitive for survey-grade applications like terrain modeling.
- **Strengths:**
    - Long range (up to 600m) supports high-altitude mapping of large areas.
    - High point rate and channel count produce dense, detailed point clouds.
    - SLAM capabilities enhance performance in GPS-denied environments.
- **Weaknesses:**
    - Subscription-based software increases long-term costs.
    - Weight (1.8 kg) requires robust drones, limiting compatibility with smaller platforms.
    - Customer service concerns reported by some users may impact support.

End-User Expectations: The Rock 3 Pro is ideal for professionals tackling large-scale, high-detail projects, such as infrastructure surveys or SLAM-based modeling. Its long range and high performance are key advantages, but users should be prepared for software costs and robust drone requirements.

Comparative Summary

Unit	Channels	Returns	Points/Sec	Weight	Cost ($K)	Resolution (cm)	Software Dependency
Quanergy M8	8	3	420,000	900g	12–15	0.3	Qortex (Subscription)
DJI Zenmuse L2	6 (~24)	5	240,000	1.1 kg	13–15	0.4	DJI Terra (Subscription)
Hesai PandarXT-32	32	3	640,000	1.1 kg	15–18	0.25	PandarView (Basic)
Ouster OS1	64	4	1,310,000	1.2 kg	20–25	0.2	Ouster Studio (Basic)
YellowScan Mapper	6 (~24)	3	240,000	1.1 kg	20–25	0.5	CloudStation (Licensed)
Phoenix LiDAR XT32	32	3	640,000	1.5 kg	30–40	0.25	LiDARMill (Subscription)
Rock 3 Pro	40	4	1,200,000	1.8 kg	35–45	0.15	SpatialExplorer (Subscription)
Teledyne Optech Maverick	32	4	800,000	1.3 kg	40–50	0.2	LMS Pro (Licensed)
Riegl miniVUX-3UAV	~100	5	200,000	1.6 kg	50–70	0.1–0.2	RiPROCESS (Complex)
Leica BLK2FLY	~32	4	700,000	2.5 kg	80–100	0.2	Cyclone (Complex)

Practical Considerations for End Users

Project Scale and Environment: For small-scale or budget-limited projects, the Quanergy M8, DJI Zenmuse L2, or Hesai PandarXT-32 offer affordable entry points, though they may struggle in dense or complex environments. For large-scale or high-precision tasks (e.g., forestry, urban mapping), the Riegl miniVUX-3UAV, Ouster OS1, Leica BLK2FLY, or Rock 3 Pro are recommended due to their superior channel counts and return capabilities.

Software Workflow: Most units rely on proprietary software for initial processing, which can be a bottleneck. Budget-conscious users should prioritize units like the Hesai PandarXT-32, where free (albeit basic) software reduces costs. Professionals with complex workflows may prefer the Riegl miniVUX-3UAV or Leica BLK2FLY, despite their software complexity, for robust integration options.

Cost vs. Resolution Trade-Off: Higher-cost units like the Riegl miniVUX-3UAV, Leica BLK2FLY, and Teledyne Optech Maverick deliver superior resolution, justifying their price for survey-grade applications. For general-purpose mapping, the Hesai PandarXT-32, Phoenix LiDAR XT32, or YellowScan Mapper offer a strong balance of cost and quality.

Drone Compatibility: Lightweight units like the Quanergy M8 and Hesai PandarXT-32 are ideal for smaller drones, while heavier units like the Leica BLK2FLY or Rock 3 Pro require robust platforms with sufficient payload and power capacity.

Environmental and Operational Needs: Units like the Quanergy M8, YellowScan Mapper, and Teledyne Optech Maverick are well-suited for harsh environments due to their durability ratings. The Leica BLK2FLY's SLAM capabilities make it unique for GPS-denied environments, while the Rock 3 Pro's long range supports high-altitude mapping.

Selecting the right LiDAR unit for drone mapping depends on project requirements, budget, and operational constraints. The Quanergy M8, DJI Zenmuse L2, and Hesai PandarXT-32 are excellent for entry-level or small-scale applications, offering affordability and ease of use. The Ouster OS1, YellowScan Mapper, and Phoenix LiDAR XT32 cater to mid-range needs with solid performance and versatility. For professional-grade, high-precision mapping, the Riegl miniVUX-3UAV, Leica BLK2FLY, Teledyne Optech Maverick, and Rock 3 Pro stand out, though they come with higher costs and complexity. By understanding each unit's strengths, weaknesses, and software dependencies, end users can make informed decisions to optimize their drone mapping workflows.

Thermal Drone Sensors
Author's Note

While thermal drones are a fascinating concept and typically top the list of "what I want to buy" when it comes to drone purchases by individuals and small drone companies, outside of very small, specialized fields, they are not generally suited for any type of mapping and have minimal use case in general drone work. Even thermal drones equipped with RTK do not generally geotag images with time-sync data on RGB images. Thus, their RTK is normally only used for real-time position data in search and rescue. The other major issue with thermal drones is that they are virtually always equipped with comparatively small RGB sensors, making them an inferior choice for RGB mapping. Most are equipped with 1/2-inch CMOS 12-megapixel rolling shutter cameras. This results in slow mapping speeds and inaccurate final products. Thermal drones have their uses, mainly in inspection and search and rescue. They are a very specialized field and will likely require an additional aerial thermography certification to be of any actual use in a business. Even then, the level of available work is an extremely small portion of what is available in the industry as a whole. For example, thermal drones represent over 14% of total capital expenditures in the author's drone business while generating less than 2% of all revenue. They have been included here so the reader understands their value and limitations. Do not make this your first drone purchase unless you have already established a specific use case. It can be a quick $7000 to $30,000 mistake.

Thermal drone sensors, commonly called infrared cameras, are advanced devices attached to drones that capture thermal imagery. These sensors detect infrared radiation emitted from objects and convert it into visual images that depict temperature variations. This technology is pivotal in scenarios where thermal patterns can indicate underlying issues, such as electrical faults, insulation failures, or hidden water leaks.

Traditionally, thermal imaging required bulky equipment and was often restricted to handheld or mounted systems that could be cumbersome and time-consuming to operate. However, with the advent of Uncooled VOx Microbolometer sensors, thermal cameras can now be manufactured much smaller, lighter, and less expensively. By integrating these newer sensors into drones, it becomes possible to conduct aerial searches or inspections over large or inaccessible areas efficiently, safely, and with high precision. This has had profound implications for various industries, particularly in maintaining and monitoring large infrastructure like power plants, solar farms, wind turbines, storage tanks, and large structures.

Thermal drones are also used in big game recovery; however, laws vary greatly on the use of drones in all aspects of hunting. If you intend to purchase a drone to start a game recovery business, call your local wildlife regulation agency first. In many states and countries, the use of drones, even to recover wounded or downed game animals, is considered poaching. Violating poaching laws can have severe consequences. As stated previously, this book is not meant as a textbook on the legalities of drone operations; call a legal professional in your area who will be able to give you the correct information.

Radiometric Thermal Imaging

Radiometric thermal imaging is a specialized technology used in inspection, mapping operations and on a limited basis is search and rescue to detect and measure heat signatures with high precision. Unlike standard thermal imaging, which visually represents heat variations in grayscale or color, radiometric thermal imaging captures and quantifies the actual temperature data of objects or surfaces in a thermal image. Each pixel in a radiometric thermal image contains precise temperature information, typically measured in degrees Celsius or Fahrenheit, allowing operators to analyze heat signatures accurately.

Radiometric thermal cameras are typically mounted on drones to identify heat sources in challenging environments, such as dense forests, urban areas, or disaster zones. For example, in search and rescue missions, like locating a missing person, these cameras can detect the body heat of a human (typically around 36-37°C) against cooler backgrounds, normally at night, even through light foliage. The radiometric data enables operators to distinguish between a living person, an animal, or an inanimate heat source, whereas, when ambient temperatures are warm, a non-radiometric sensor is more likely to be washed out by the surrounding environment, thus, RMT sensors can greatly improve decision-making. The technology's larger applications like thermal inspections of infrastructure such as solar or wind farms by providing precise temperature measurements over large areas or hard to reach places. Its ability to function in low-visibility conditions, such as darkness or smoke, makes it invaluable for time-critical operations.

Advantages of Using Thermal-Equipped Drones

Drones equipped with thermal cameras bring several advantages to industrial inspections and surveillance:

- **Efficiency**: Drones can quickly cover large areas that would otherwise take hours or even days to inspect manually, significantly speeding up the maintenance, inspection, or search process.

- **Safety**: Using drones minimizes the need for personnel to physically access hazardous or hard-to-reach areas, thereby reducing the risk of accidents and injuries.

- **Accuracy**: Thermal drones provide detailed images that help identify problems like leaks, overheating, or structural weaknesses with great precision. This can be crucial in preventing failures and ensuring system reliability.

- **Cost-effectiveness**: Thermal drones reduce the reliance on expensive equipment such as cranes, helicopters, or extensive scaffolding setups, thereby cutting operational costs significantly.

Power Plants Inspection

Power plants feature a variety of critical components, such as boilers, steam turbines, and cooling towers, each with unique operational challenges. Traditional inspection techniques involve substantial downtime and pose risks due to the high-temperature and high-pressure environments common in power plants.

Thermal-equipped drones offer a powerful alternative, allowing real-time surveillance of these components without interrupting plant operations. They can detect irregular heat signatures that indicate potential issues like overheating, wear and tear on moving parts, or blockages in cooling passages. Identifying and addressing these issues promptly can prevent costly breakdowns and enhance the operational efficiency of power plants.

Solar Farm Inspection

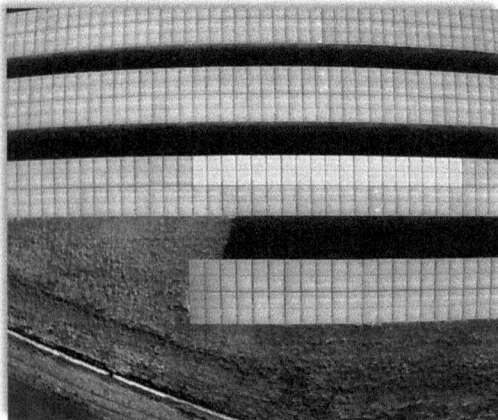

Solar farms consist of vast arrays of photovoltaic cells that can be compromised by defects or environmental factors. Inspecting each panel individually is labor-intensive and ineffective, especially in larger installations.

Using thermal drones, operators can quickly identify panels that exhibit abnormal heat patterns indicative of malfunctions such as electrical faults or inefficiencies. This helps maintain optimal energy production and extends the lifespan of solar panels by ensuring timely maintenance and replacement of defective components.

Wind Mill Inspection

Wind turbines, with their towering heights and remote locations, are particularly challenging to inspect manually. Key components such as gearboxes, blades, and generators need regular monitoring to detect signs of wear or damage.

Thermal imaging via drones can reveal issues like overheating in gearboxes or friction points along the blades, which might go unnoticed during visual inspections. By catching these early signs, maintenance teams can schedule repairs before more serious damage occurs, reducing downtime and maintenance costs.

Storage Tanks Inspection

Storage tanks, used widely in industries like oil and gas, chemicals, and water treatment, often contain substances at high temperatures or that are chemically aggressive. Leakages or insulation failures can lead to significant safety hazards and financial losses.

Thermal drones can safely assess these tanks from a distance, pinpointing areas of heat loss or unusual thermal signatures that suggest potential leaks or weaknesses in the tank structure. This proactive detection plays a crucial role in averting environmental disasters and ensuring the integrity of storage facilities.

Roof Inspection

Roofs are susceptible to a range of issues, from water leaks to poor insulation or structural damage due to environmental stressors. Traditional roof inspections can be slow, risky, and often imprecise.

Thermal-equipped drones revolutionize this task by providing a bird's-eye view of the roof's condition, highlighting areas of heat loss or moisture penetration without the need for direct contact. This method is faster, safer, and offers a more comprehensive analysis of the roof's condition, facilitating more accurate maintenance and repair.

Integrating thermal sensors in drone technology has transformed how infrastructures are inspected and maintained. With their ability to rapidly and safely collect detailed thermal data, these drones are invaluable in enhancing maintenance programs' operational efficiency, safety, and cost-effectiveness across various industries. As technology advances, the scope and capabilities of thermal-equipped drones will expand, promising even more significant impacts on industrial health monitoring and preventative maintenance strategies.

Thermal Pet Search and Rescue

In scenarios where pets go missing, especially in expansive or challenging terrains, traditional search methods can be time-consuming and often less effective. Thermal-equipped drones provide a significant advantage in these situations, enabling search teams to cover large areas quickly and detect heat signatures that are indicative of a hidden or stranded animal.

Use Case

Thermal drones are deployed in search and rescue operations for missing pets, particularly in dense forests or during nighttime when visibility is low. The thermal sensors can easily identify the heat emitted by animals, distinguishing them from the cooler surroundings. This capability is crucial in promptly locating pets that are scared, injured, or trapped, significantly increasing the chances of a successful rescue. By utilizing thermal imaging, rescuers can minimize the search time and ensure the safety of the pet, even in adverse conditions such as cold weather or under foliage where visual detection is challenging.

Thermal Big Game Recovery

Big game hunting often involves tracking injured game across difficult terrains, which can be both challenging and time-consuming. Thermal imaging technology equipped on drones can transform this task by pinpointing the location of game based on their heat signatures, particularly during night hunts or in dense brush.

Use Case

After an animal has been hit, a thermal-equipped drone can be deployed to track its path. The drone flies over the area where the game was last seen, using thermal sensors to detect the heat emitted by the animal's body. This method is especially effective in tracking wounded game that has retreated into thickets or other obscure places. By providing precise location data, the drone helps hunters or wildlife managers recover the animal swiftly, reducing the animal's suffering and ensuring that the game is handled ethically and responsibly. This approach not only enhances the efficiency of the recovery process but also supports conservation efforts by ensuring that wounded animals are quickly located and not left to waste.

Thermal Drones

The Teledyne FLIR Siras

The Teledyne FLIR Siras is a cutting-edge drone designed for professional use in various industries, including public safety, inspection, and environmental monitoring. The drone is equipped with a modular payload system and engineered to provide high-quality thermal imaging and precise data collection in challenging environments. Currently, only the FLIR Boson 640x512 thermal payload is available.

- **Key Specifications**

    - **Weight**: Approximately 3.2 kg, making it light enough for easy handling yet robust enough for stable flight.

    - **Payload**: Equipped to handle various sensors, including the integrated Teledyne FLIR radiometric thermal camera and a visible light camera.

    - **Flight Time**: Up to 31 minutes with a fully charged battery, suitable for short to medium-range missions.

- **Features**

    - **Integrated Thermal and Visual Cameras**: The drone features a dual camera setup with a Teledyne FLIR radiometric thermal sensor and a high-resolution visible light camera. This combination allows for the simultaneous capture of thermal and standard imagery, facilitating detailed inspections and situational awareness.

    - **Advanced Image Processing**: Includes onboard image processing capabilities that enhance image quality and accuracy, making it easier to interpret thermal data directly from the drone.

    - **User-Friendly Operation**: The Teledyne FLIR Siras is designed for ease of use. It features automated flight modes, simple control interfaces, and integrated mission planning software.

    - **Data Integration and Analysis**: Compatible with various data analysis and management platforms, allowing for easy integration of thermal data into existing workflows for further analysis and decision-making.

    - **Rugged Design**: Built to withstand challenging weather conditions, ensuring reliable operation in diverse environments.

The Teledyne FLIR Siras thermal drone stands out for its integration of high-quality thermal imaging capabilities with the ease and flexibility of a drone platform. This makes it an invaluable tool for professionals needing rapid, reliable thermal assessments in fields such as emergency response, industrial inspections, and environmental monitoring.

Skydio X10 VT300-L

The Skydio X10 is a sophisticated unmanned aerial vehicle that integrates advanced AI-driven technologies with robust hardware for high-performance applications. It is purpose-built to deliver superior imaging and data collection capabilities, especially for demanding environments.

- **Key Specifications**
    - **Weight**: Approximately 1.7 kg, designed for optimal balance between durability and agility.
    - **Dimensions**: Compact and portable, facilitating ease of transport and rapid deployment.
    - **Payload**: This device features a dual-camera system with a "narrow" 64MP color camera, a wide 1-inch CMOS 50MP color camera, and a 640x512 radiometric thermal imager, providing comprehensive visual and thermal data capture.
    - **Flight Time**: Up to 35 minutes, allowing for extensive operational periods and detailed area coverage.
    - **Maximum Speed**: Capable of speeds up to 36 km/h, ensuring efficient navigation across various terrains.
    - **Operational Range**: Up to 6.2 km, providing a broad reach for extended missions.

- **Features**
    - **Advanced Imaging Capabilities**: The integrated cameras offer high-resolution visual and thermal imaging, enabling users to capture detailed, actionable data for various applications, from infrastructure inspections to emergency response.
    - **AI Autonomy**: Skydio X10 leverages Skydio's proprietary AI technology for autonomous navigation, which reduces pilot workload and enhances safety by avoiding obstacles even in complex environments.
    - **Real-Time Data Processing**: Equipped with advanced processing capabilities that facilitate immediate data analysis, aiding in quick decision-making and efficient workflow management.
    - **Robust Design**: Constructed to withstand challenging operational conditions, the drone is weather-resistant and durable, suitable for rigorous field use.
    - **Seamless Integration**: Compatible with various data management and analysis platforms, the Skydio X10 can easily integrate into existing workflows, enhancing data utility and accessibility.

The Skydio X10 is a good tool for professionals requiring a reliable and precise aerial solution, particularly in sectors requiring detailed environmental monitoring, rapid response, and critical data collection. Its combination of advanced imaging, robust autonomous capabilities, and durable construction makes it an invaluable asset in enhancing operational effectiveness and safety.

Autel Robotics Evo Max 4T

The Autel Robotics Evo Max 4T is a versatile drone designed for professional use, particularly in public safety and inspection. It is distinguished by its multi-sensor payload and robust capabilities, making it ideal for complex operational demands.

- **Key Specifications**

  - **Payload**: Equipped with an integrated triple-sensor payload that includes a thermal camera, a high-resolution RGB camera, and a laser rangefinder.

  - **Flight Time**: Up to 42 minutes on a single battery charge, enabling extended operations over large areas.

  - **Cruise Speed**: Capable of reaching speeds up to 27 m/s, allowing for rapid transit and response.

  - **Maximum Operating Range**: Up to 18.5 km (12 miles) with robust anti-interference capabilities, ensuring reliable control at long distances.

  - **Operating Temperature**: Functions in environments ranging from -10°C to 40°C, suitable for various climatic conditions.

- **Features**

  - **Triple-Sensor System**: This system includes a 640x512 thermal camera, a 50 MP RGB camera, and a laser rangefinder with a range of up to 1200 meters. This combination allows for versatile data collection in a single flight.

  - **Autonomy and Intelligence**: Features advanced obstacle avoidance systems and intelligent flight modes, including autonomous patrol, smart tracking, and precision waypoint flying.

  - **High Precision Positioning**: Equipped with RTK (Real-Time Kinematic) GPS for centimeter-level positioning accuracy, which is crucial for detailed surveying and mapping tasks.

  - **Data Security**: Offers data encryption and secure transmission features, ensuring sensitive information remains protected during UAV operations.

  - **Rugged Design**: Built to be weather-resistant, allowing for operation in diverse and challenging environments without compromising performance.

  - **User Interface and Control**: Comes with an intuitive controller and user-friendly app interface, simplifying complex operations and enhancing usability for all pilot levels.

The Autel Robotics Evo Max 4T stands out in the market for its multi-sensor payload and extensive flight capabilities. It is designed to deliver high-quality data and robust performance across various professional applications. Its advanced features and operational flexibility make it invaluable for critical missions and detailed environmental monitoring.

DJI Matrice 30T

The DJI Matrice 30T is a highly versatile and robust commercial drone designed for use in a wide variety of industrial and safety applications. These include power system inspections, public safety, search and rescue, and more. It is part of DJI's renowned Matrice series, which is known for its advanced technology and durability.

- **Key Specifications**

  o **Payload**: Equipped with an integrated payload that includes a 512x640 thermal camera, 1/2" CMOS 48MP zoom camera, a 1/2" CMOS 12MP wide-angle camera, and a laser rangefinder with a 3–1,200 m range. The zoom camera on the M30T is capable of 5x–16x optical and 200x hybrid digital zoom.

  o **Flight Time**: Up to 41 minutes, allowing for extended missions and comprehensive area coverage.

  o **Maximum Speed**: Can reach speeds up to 82 km/h, enabling rapid response and efficient area traversal.

  o **Operating Range**: Up to 15 km with OcuSync 3 Enterprise for stable, reliable HD image transmission.

  o **Maximum Altitude**: Capable of flying up to 7000 meters above sea level, though operational ceiling might be lower depending on local regulations.

  o **Operating Temperature**: Built to operate between -20°C to 50°C, making it suitable for a variety of harsh environments.

- **Features**

  o **Triple-Sensor Payload**: Includes a 48 MP zoom camera with a 200x zoom, a 12 MP wide-angle camera, and a 640x512 px radiometric thermal camera, providing versatile imaging options for detailed inspections and situational awareness.

  o **Advanced Flight Modes**: Supports multiple intelligent flight modes, including waypoint, orbit, and one-key takeoff/landing, which enhance operational efficiency and safety.

  o **Robust Build Quality**: Features a rugged design with an IP55 rating for resistance to dust and water, ensuring reliability in adverse weather conditions.

  o **Safety and Navigation**: Integrated with ADS-B receivers to enhance airspace safety by alerting operators of nearby manned aircraft.

  o **Autonomous Features**: AI capabilities for smart-tracking and obstacle avoidance ensure smooth, safe operations in complex environments.

- **Data Security**: Comes equipped with local data mode and onboard storage, minimizing the risk of data breaches.

The DJI Matrice 30T is tailored for professionals who demand a reliable, high-performance UAV that can handle demanding tasks across diverse operational scenarios. Its combination of powerful imaging capabilities, extended flight time, and robust construction make it an invaluable tool for enhancing productivity and safety in a wide range of industrial applications. A unique feature of this drone is its ability to track targets from up to two miles away. Unlike the "follow me" option found on some drones, this system will remain stationary while locking on and following targets with its zoom camera. The M30T is also equipped with a 1920x1080 FPV camera strictly for pilot operation. This feature is included on the M30T because it can connect two remotes simultaneously. This ability can be utilized in two different ways. The first is the ability to hand off control while in flight. At any given time, one pilot can hand off control to another. The second option allows the second operator to fly the gimbal independently of the pilot for search and rescue operations. Lastly, when the drone is in search and rescue mode, the controller can be connected directly to an internet connection via an optional 4G modem or a hotspot and broadcast a live feed of the captured data via FlightHub 2.

The series of thermal drones we have discussed so far all have a stated weather resistance rating (IPX rating), making them extremely useful in situations where weather is a factor, such as search and rescue (SAR). Alongside these larger, weather-resistant drones, a series of thermal drones have entered the market that are smaller, have a few fewer features, and lack a weather resistance rating. This does not, however, make them incapable of performing most tasks. While they are not as useful for all-weather missions, they are extremely capable in the field of thermal inspection. Their compact, single-battery designs make them easy to deploy, faster into the air, and less intrusive in populated areas. These little "mighty mites" are well suited for inspection of solar farms, windmills, roofs, and many other forms of inspection. Some factories also use drones such as these, fitted with prop guards, for interior inspections.

Compact Thermal Drones

DJI Mavic 3 Thermal / Anzu Robotics Raptor T*

The DJI Mavic 3 Thermal is a high-performance drone designed for a wide range of applications, including infrastructure inspection, environmental monitoring, and wildlife management. This drone combines the advanced portability and ease of use that the Mavic series is known for with enhanced thermal imaging capabilities, making it a powerful tool for professionals requiring visual and thermal data capture.

As with the Anzu Robotics Raptor and DJI Mavic 3 Enterprise, the Anzu Robotics Raptor T is a licensed copy of the Mavic 3 Thermal, built in a neutral country using all American-written software.

- **Key Specifications**

  - **Weight**: Approximately 899 grams, making it lightweight and highly portable.

  - **Flight Time**: Up to 46 minutes, providing extended periods for missions and reducing the need for frequent landings to change batteries.

  - **Operating Range**: OcuSync technology offers a transmission range of up to 15 km, ensuring a robust and reliable connection even at long distances.

- **Features**

  - **Advanced Imaging System**: Features a 1/2-inch CMOS 48MP wide camera and 1/2-inch CMOS 12MP zoom for capturing photos and videos alongside a 512x640 thermal imaging camera that provides valuable thermal data for inspections and emergency response.

  - **Enhanced Autonomy**: Equipped with Advanced Pilot Assistance Systems (APAS) 5.0, which helps the drone autonomously avoid obstacles in complex environments.

  - **Data Security**: Incorporates robust data security measures to protect sensitive information suitable for professional use in critical sectors.

The DJI Mavic 3 Thermal is an outstanding drone for professionals needing a reliable, versatile UAV that provides detailed visual and thermal imagery. Its robust feature set ensures it can perform in a range of conditions and scenarios, making it a preferred choice for critical and precise aerial data collection.

Autel Evo 2 Enterprise Dual 640T V3*

The Autel Evo 2 Dual 640T Enterprise V3 is an advanced drone designed for professional applications, including public safety, emergency response, infrastructure inspections, and environmental monitoring. This model is part of the Autel Robotics Evo series, known for its robust performance and high-quality imaging capabilities, and comes equipped with an enhanced thermal imaging system.

As denoted by the "V3," this is the third version of the Autel Evo 2 thermal drone. It is important to note that each version, while identical in appearance, is not compatible with other versions internally. Due to supply chain issues, Autel was forced to change some hardware, creating incompatibility and different versions. While batteries, RTK modules, and accessories such as external speakers and spotlights are interchangeable, controllers and internal components are not. Unlike virtually every other drone in this book, the V2 and V1 versions of this drone are NOT Remote ID compliant and will require a separate Remote ID module to operate legally.

- **Key Specifications**

  - **Weight**: Approximately 1.174 kg (2.59 lbs), providing a solid balance between durability and maneuverability.

  - **Payload**: This device features a triple sensor array with a high-resolution visible camera, an infrared thermal camera, and a laser rangefinder.

  - **Flight Time**: Up to 40 minutes, allowing for extended operational periods and comprehensive area coverage.

  - **Operating Range**: Offers a robust communication range of up to 9 km (5.6 miles) with a stable HD transmission system.

- **Features**

  - **Advanced Imaging Capabilities**: Equipped with a 640 x 512 thermal imaging sensor and a 50 MP visible light camera, providing detailed data capture for dual imaging in both thermal and visible spectrums.

- **Obstacle Avoidance**: Integrates Autel's proprietary obstacle avoidance technology, ensuring safe navigation in complex environments.

- **Autonomous Flight Modes**: Supports multiple intelligent flight modes, such as Dynamic Track, Orbit, and Dual Stability, which enhance data collection and operational efficiency.

- **Robust Data Security**: Offers secure data transmission and storage options, essential for sensitive and critical missions.

- **Modular Accessories**: Compatible with additional modular accessories like loudspeakers, spotlights, and beacons, which expand its utility in various professional contexts.

The Autel Evo 2 640T Enterprise V3 is a highly capable and versatile drone built to meet the demanding needs of professionals in the field. Its sophisticated sensors and modular design make it invaluable for achieving precise and efficient results in a wide range of industrial and emergency scenarios.

Hyperspectral Sensors

Hyperspectral sensors represent a significant advancement in remote sensing technology. They capture images across hundreds of narrow, contiguous spectral bands. Unlike traditional RGB sensors, which capture data in three broad color bands, and multispectral cameras, which capture images in five to six bands, hyperspectral sensors provide detailed spectral information that allows for precise material and condition analysis of the Earth's surface across a very wide electromagnetic spectrum.

It's important to note that, unlike multispectral sensors, which consist of 5–6 individual sensors set to capture particular wavelengths, hyperspectral sensors can vary greatly in their design, sensing technology, and abilities. While they are grouped in one section for the benefit of a basic explanation, they are actually a very diverse set of different types of sensors that may perform very different functions. Some use lasers, some use air intake sensors, and some use specially designed cameras. This section is meant as a general overview of their uses and abilities, not a complete guide. If you choose to enter the world of hyperspectral inspection, please spend the time to research exactly what you will need to perform the mission you require. Most of these sensors will cost in excess of $100,000 to purchase, not including a drone.

How Hyperspectral Sensors Work

Hyperspectral imaging works by collecting and processing information from the electromagnetic spectrum. Each object has a unique spectral signature based on how it reflects light across different wavelengths. By capturing these signatures, hyperspectral sensors can accurately identify and differentiate between various materials and conditions.

Applications of Hyperspectral Sensors

- **Agriculture**: These sensors can also analyze the spectral signatures of different crops and their conditions to provide precise crop health monitoring, disease detection, and nutrient level assessment. Because they detect all of the same bands as multispectral cameras, they are virtually interchangeable in this respect. However, due to their cost-prohibitive nature, they are not the first choice for crop monitoring. Standard multispectral sensors are a fraction of the cost and provide virtually all the data needed for standard agriculture.

- **Environmental Monitoring**: They are effective in detecting pollutants in water, monitoring vegetation health, and mapping biodiversity.

- **Mining and Geology**: Hyperspectral sensors can identify minerals and rock types, map geological formations, and aid resource exploration.

- **Forestry**: They assist in forest management by assessing tree health, species identification, and tracking deforestation.

- **Forensic and Security**: These sensors help in detecting specific materials or chemicals, aiding forensic investigations or environmental monitoring.

Advantages of Hyperspectral Imaging

- **Detailed Analysis**: Provides comprehensive information about the physical and chemical properties of materials.

- **Material Identification**: Can accurately identify and differentiate between specific materials and substances.

- **Versatile Applications**: Useful across a wide range of industries, including agriculture, environmental science, mining, and security.

Best Practices and Considerations

- **Cost and Complexity**: Hyperspectral imaging systems are more expensive and complex than RGB or multispectral cameras, requiring specialized data processing and analysis knowledge.

- **Data Processing**: The vast amount of data collected necessitates robust data processing systems and software.

- **Operational Challenges**: Planning and executing hyperspectral imaging missions can be more challenging due to their complexity and data requirements.

Hyperspectral sensors offer unparalleled detail and precision in remote sensing, making them invaluable for projects requiring detailed material analysis. Despite the higher cost and complexity, the depth and quality of data provided by hyperspectral imaging can offer significant benefits across various applications.

Gas Detection with Hyperspectral Sensors for Drones

Introduction to Gas Detection

Hyperspectral sensors are increasingly being utilized for gas detection applications due to their ability to capture detailed spectral information across a wide range of wavelengths. This capability allows for identifying and quantifying various gases based on their unique spectral signatures.

How Hyperspectral Sensors Detect Gases

Hyperspectral sensors detect gases by capturing the specific wavelengths of light absorbed and emitted by gas molecules. Each type of gas has a distinct spectral signature that can be identified by analyzing the hyperspectral data. This process involves:

- **Spectral Signature Analysis**: Gases absorb light at specific wavelengths. Hyperspectral sensors can detect and identify the presence of different gases by measuring the absorption at these wavelengths.

- **Remote Sensing**: Drones equipped with hyperspectral sensors can survey large areas from the air, providing a comprehensive view of gas distributions and concentrations.

Applications of Gas Detection with Hyperspectral Sensors

- **Environmental Monitoring**: Detection of greenhouse gases such as methane (CH_4) and carbon dioxide (CO_2) to monitor pollution levels and assess the impact of industrial activities on the environment.

- **Industrial Safety**: Monitoring for leaks of hazardous gases in industrial facilities, ensuring safety and compliance with environmental regulations.

- **Agriculture**: Detection of gases emitted from soil and crops can indicate soil health and crop conditions.

- **Disaster Response**: Identifying hazardous gas releases during natural disasters such as wildfires or chemical spills, aiding in emergency response efforts.

Advantages of Hyperspectral Gas Detection

- **High Sensitivity**: Capable of detecting low concentrations of gases with high precision.

- **Wide Area Coverage**: Drones can cover large areas quickly, providing extensive data on gas distributions.

- **Non-Invasive**: Remote sensing with drones is non-invasive, causing no disturbance to the environment or industrial processes being monitored.

Case Studies and Real-world Applications

- **Methane Detection**: Hyperspectral sensors have been used to detect methane leaks in oil and gas facilities, providing critical data for leak detection and repair programs.

- **Greenhouse Gas Monitoring**: Drones equipped with hyperspectral sensors have been deployed to monitor greenhouse gas emissions in urban areas, contributing to climate change studies and mitigation efforts.

Challenges and Considerations

- **Data Complexity**: The data collected from hyperspectral sensors is complex and requires advanced processing techniques to extract meaningful information.

- **Cost**: Hyperspectral gas detection systems are generally more expensive than other gas detection technologies, though the detailed information they provide can justify the investment for critical applications.

Hyperspectral sensors offer a powerful tool for gas detection, providing high sensitivity and the ability to cover large areas quickly. Their applications in environmental monitoring, industrial safety, agriculture, and disaster response highlight their versatility and importance in modern remote sensing efforts. Understanding the capabilities and applications of hyperspectral sensors can help leverage this advanced technology for various gas detection needs.

Mineral Detection with Hyperspectral Sensors for Drones

Hyperspectral sensors are increasingly utilized in geological and mining applications for their ability to detect and identify various minerals. Hyperspectral sensors can provide precise identification and mapping of mineral compositions by capturing detailed spectral information across a wide range of wavelengths.

How Hyperspectral Sensors Detect Minerals

Mineral detection using hyperspectral sensors involves analyzing the unique spectral signatures of different minerals. Each mineral reflects and absorbs light differently across the electromagnetic spectrum. Hyperspectral sensors can identify and differentiate between various minerals by capturing these variations.

- **Spectral Signature Analysis**: Each mineral has a unique spectral signature. By analyzing the light reflected from a mineral surface across multiple wavelengths, hyperspectral sensors can identify the specific minerals present.

- **Remote Sensing**: Drones equipped with hyperspectral sensors can cover large and inaccessible areas, providing comprehensive mineral maps from the air.

Applications of Mineral Detection with Hyperspectral Sensors

- **Mining Exploration**: Identifying and mapping mineral deposits, assisting in the exploration and assessment of mining sites.

- **Geological Mapping**: Detailed geological surveys to understand the composition and structure of the Earth's surface.

- **Environmental Monitoring**: Assessing the environmental impact of mining activities by detecting changes in soil and vegetation composition.

- **Archaeological Studies**: Detecting mineral-rich artifacts and features in archaeological sites.

Advantages of Hyperspectral Mineral Detection

- **High Precision**: Capable of identifying minerals with high accuracy due to the detailed spectral information.

- **Wide Area Coverage**: Drones can efficiently cover large and remote areas, providing extensive mineral data.

- **Non-Destructive**: Remote sensing with hyperspectral imaging is non-invasive, preserving the integrity of the survey area.

Case Studies and Real-world Applications

- **Copper and Iron Ore Detection**: Hyperspectral sensors have been used to map copper and iron ore deposits, providing valuable data for mining companies to optimize extraction processes.

- **Rare Earth Element Exploration**: Drones equipped with hyperspectral sensors have been deployed to identify rare earth elements, which are crucial for high-tech industries.

Challenges and Considerations

- **Data Complexity**: The data collected from hyperspectral sensors is complex and requires advanced processing techniques to extract meaningful mineral information.

- **Cost**: Hyperspectral mineral detection systems are generally more expensive than traditional geological survey methods, though the detailed information they provide can justify the investment for critical applications.

Popular Hyperspectral Sensors for Drones

- **Specim AFX Series**: Known for its compact, all-in-one design, the Specim AFX series integrates a high-end hyperspectral sensor, a powerful onboard computer, and a GNSS/IMU unit. It is suitable for various applications, including vegetation classification, water quality analysis, and plant health studies.

- **Resonon Pika L and Pika IR-L**: These sensors are lightweight and compact, making them ideal for UAVs. They cover the visible to near-infrared spectral ranges and are used for precision farming, environmental monitoring, and geological exploration.

Hyperspectral sensors offer a powerful tool for mineral detection, providing high precision and the ability to cover large areas efficiently. Their applications in mining exploration, geological mapping, and environmental monitoring highlight their importance in modern geological studies. Understanding the capabilities and applications of hyperspectral sensors can help leverage this advanced technology for various mineral detection needs.

Chapter 5: GNSS and Positioning Systems

This chapter delves into the essential role of Global Navigation Satellite Systems (GNSS) and positioning technologies in drone mapping, with GNSS encompassing systems like the U.S.-maintained GPS and others such as GLONASS, Galileo, BeiDou, NavIC, and QZSS. It explores the intricacies of GPS, RTK (Real-Time Kinematic), and PPK (Post-Processing Kinematic) systems. The chapter highlights the significance of precise positioning for accurate geospatial data collection and explains how various systems like CORS networks, local base stations, and Ground Control Points (GCPs) enhance this accuracy. Additionally, it examines the integration of RTK and PPK technologies, showcasing their practical applications across different sectors. It also introduces advanced tools like Propeller AeroPoints that further streamline and improve the reliability of drone mapping operations.

Accurate positioning is crucial in drone mapping. This is achieved through various technologies:

- **GPS**: GPS is a global navigation satellite system maintained by the United States government and freely accessible by anyone with a GPS receiver. The system provides geolocation and time information to a GPS receiver anywhere on Earth with an unobstructed line of sight to four or more GPS satellites. It relies on a constellation of satellites that transmit precise signals, enabling GPS receivers to determine their location, speed, direction, and time. In addition to the American GPS system, several other countries operate similar constellations. These include GLONASS (Russian Federation), Galileo (European Union), BeiDou (China), NavIC (India), and QZSS (Japan).

- **RTK (Real-Time Kinematic)**: RTK enhances GPS accuracy by correcting signal errors in real time using a fixed reference station. These fixed reference stations can be CORS or local base stations, as discussed later in this section. RTK is crucial for applications requiring centimeter-level precision in measurements.

- **PPK (Post-Processing Kinematic)**: Similar to RTK, PPK corrects GPS data but does so after the flight rather than in real time. This method is useful when real-time data transmission is not feasible or when local accuracy may deviate from global accuracy.

- **CORS (Continuously Operating Reference Station)**: CORS networks support RTK and PPK operations by providing additional data points (towers) for error correction. They enhance the accuracy of position data across larger regions. There are both public and private CORS networks across the United States. Many are operated by state highway departments, others by private companies. Still others are crowdfunded and are growing as individuals work together to assemble networks specifically for the drone industry. When looking for a CORS network in your area, it is best to check with your highway department first. In many states, they are free and highly reliable.

Real-Time Kinematic (RTK) Technology in Drone Mapping
Real-Time Kinematic (RTK) technology represents a significant advancement in precision navigation and positioning systems. It enhances GPS data to provide centimeter-level accuracy. This high precision is crucial in drone mapping, where detailed and accurate geospatial data is necessary. This section delves into the workings of RTK GPS, its essential equipment, and its critical applications in the field.

Technology Overview

RTK GNSS is a type of differential GPS that enhances the precision of position data derived from satellite-based positioning systems. It operates using a fixed base station that broadcasts the difference between the positions indicated by the satellite systems and the known fixed position. This correction signal is then received by a mobile receiver, which is used by the drone to correct its position. By correcting these signals in real time, RTK GNSS can achieve positioning accuracy down to the centimeter level. This significantly refines the accuracy compared to uncorrected GPS, which can have several meters of error.

Integrating RTK technology in drones revolutionizes how they perform tasks requiring high precision, such as surveying, mapping, and inspection. Drones equipped with RTK GNSS can navigate more accurately, maintain a stable position, and collect geospatial data with high reliability. This holds true even under challenging conditions or in environments with obstructions that typically interfere with traditional GPS signals. As listed below, an RTK base station or a direct connection to a CORS network is required for accurate RTK corrections. In cases where extreme accuracy is needed, PPK and RTK may be employed.

Equipment Necessary for RTK to Function

Implementing RTK technology involves specific hardware components that ensure its functionality:

- **Continuously Operating Reference Stations (CORS) Networks**: CORS networks consist of permanently installed GNSS receivers (towers) that continuously collect satellite data. This data is used to compute high-accuracy correction information, normally synced by cellular internet connection (NTRIP) to the ground control station. This allows for precise triangulation of the drone in real time, which is essential for accurate geolocation. The correction data from CORS networks can be accessed in real time or post-processed to improve the positional accuracy of GPS systems from meter-level to centimeter-level precision.

- **RTK Base Station**: The base station is a ground-based GPS receiver that provides real-time corrections. Its location is precisely known. It continuously monitors satellite signals to calculate correction data. The base station then transmits these corrections to the drone, which uses them to adjust its GPS signals. If the base station operates alone, it can be placed on a known point set by an engineer or surveyor or use a technique called "averaging."

- **Averaging**: Averaging is a technique by which the base station continually updates its location by averaging GPS signals it receives and refining that information. Due to ever-present interference, this is the least accurate means of georeference and may be off by 1 to several meters, making it less suitable if absolute accuracy is required.

- **Known Point**: In many cases, a known point may be the preferred method of RTK. This is because, when using a known ground point, all data will be adjusted to that fixed practical known point. This will provide the best georeferenced relative accuracy. When using a GPS-derived point, while the data may be triangulated even more accurately within a geoid and ellipsoid dataset, local accuracy may be affected based on previous inaccuracies in historical surveys. It's important to remember that, while GNSS is a relatively new technology, surveying is a millennia-old art. For centuries, it depended on chain lines, hand measuring, and even dead reckoning. Many monuments still used today have never been geospatially referenced. If they are your fixed point of reference, it is best to use the data provided as the fixed point to keep all local data accurate with reference to that point. Wars have been fought over less!

- **CORS / NTRIP**: CORS is included as both a stand-alone reference and part of the base station reference because it can be connected to either the base station or directly to the drone. RTK-enabled drones have the ability to connect directly to CORS networks via NTRIP protocol or to a base station that receives an NTRIP signal (an internet connection is required when using an NTRIP signal). Because base stations are stationary and have longer duration triangulation averaging, even when connected to an NTRIP signal, they tend to be much more accurate than using an NTRIP signal connected directly to a drone. Most drones advertise an accuracy of 1 cm + 1 PPM horizontal and 1.5 cm + 1 PPM vertical. Most base stations can achieve straight centimeter-level accuracy. Please reference your hardware manual to determine the projected accuracy.

- **RTK Module on the Drone**: This module receives corrections from the RTK base station. Modern drones integrate these modules into their systems. This allows them to process correction data in real time to achieve precise positional accuracy.

- **Communication Link**: A robust communication system is crucial to ensure the uninterrupted transmission of correction data from the RTK base station to the drone. Depending on the drone's range and operational environment, this is typically accomplished via radio or cellular networks.

- **Software for Data Processing**: Besides hardware, specialized software is needed to process the data collected using RTK technology. This software interprets the corrected positional data alongside the drone's sensor data (like images or LiDAR). It aligns all information to produce highly accurate maps and models.

Applications of RTK in Drone Mapping

RTK technology is utilized in various critical applications, demonstrating its versatility and importance:

- **Detailed Site Surveys**: Precise site surveys are essential for construction and engineering projects. Drones equipped with RTK technology can perform these surveys quickly and accurately. They provide detailed topographical data crucial for planning and decision-making processes.

- **Infrastructure Inspections**: RTK drones are extensively used to inspect infrastructure such as bridges, towers, and power lines. Their ability to pinpoint exact locations and maintain stable flight patterns, even close to structures, allows for detailed and safe inspections.

- **Precision Agriculture**: In agriculture, RTK technology helps create detailed farm maps for soil sampling, drainage planning, and crop monitoring. This precision supports farmers in implementing more efficient planting, watering, and treatment plans. These can lead to increased crop yields and reduced waste.

- **Environmental Monitoring**: RTK drones are invaluable in environmental conservation projects, such as monitoring erosion, tracking habitat changes, and mapping flood-risk areas. Their precision ensures that environmental changes are accurately tracked over time. This provides reliable data for environmental assessments and regulatory compliance.

- **Crop Monitoring and Treatment**: Using RTK-equipped drones allows the farmer to isolate drought, disease, infestation, and other overstress conditions down to a single plant. While isolating a single plant is not necessarily cost-effective, knowing exactly where an issue starts and ends can help the farmer address the problem and proactively track the source.

Real-time kinematic technology enhances drones' capabilities by enabling them to achieve unprecedented accuracy in positioning and navigation. As this technology continues to evolve and integrate more seamlessly with various drone systems, its applications are expected to expand further. RTK's impact on drone mapping and other precision-dependent fields underscores its growing importance in the technological landscape. It marks it as a critical tool in modern geospatial data collection.

RTK Workflow

Real-Time Kinematic (RTK) workflow in drones involves using the CORS network connected via NTRIP connection. The process starts with the CORS network, which provides the precise GNSS-corrected data. This data is transmitted over the internet via an NTRIP server. The NTRIP server streams correction data via protocols like RTCM 3.2, which the drone's RTK module processes instantly. This requires a stable cellular or Wi-Fi connection, typically provided by a ground station or a hotspot paired with the controller. An RTK-enabled drone with an RTK GNSS receiver connects to the NTRIP server using a mobile device or ground station with internet access. It receives real-time correction data. The drone combines this correction data with its own GNSS data to achieve centimeter-level positional accuracy during flight. This is crucial for applications like surveying, mapping, and construction monitoring.

Post-Processing Kinematic (PPK) in Drone Mapping

Post-Processing Kinematic (PPK) technology, often discussed in tandem with Real-Time Kinematic (RTK), is a powerful positioning tool used in drone mapping. It processes data after the fact rather than in real time. This section explores the nuances of PPK technology, its relationship with RTK, and practical applications where PPK offers significant advantages. These are especially notable in scenarios with limited real-time data transmission capabilities.

Technology Overview

PPK, like RTK, is a form of differential GPS that enhances the accuracy of satellite data. The primary distinction lies in the timing of the corrections applied to the GPS data. While RTK corrects GPS data in real time, PPK stores all raw satellite data (from both the drone and a ground station) to be processed after the flight. This post-mission correction aligns the in-flight data with the base station data. It calculates the most accurate position for each recorded point.

One key advantage of PPK over RTK is its data handling and correction flexibility. Since PPK does not require a continuous live connection to corrected data, it is less vulnerable to data loss due to signal interruptions. These can be caused by topographical obstructions, radio interference, or distance limitations from the base station.

Relationship with RTK

Both PPK and RTK are based on the principle of using a base station that records GPS errors, which are crucial for correcting positional inaccuracies. Here are the main differences and considerations for choosing between the two:

- **Data Reliability**: PPK generally offers higher reliability in data collection because it does not depend on a live connection. This makes it particularly useful in remote or challenging environments where maintaining a stable real-time link can be difficult.

- **Operational Flexibility**: RTK requires a real-time link between the drone and the base station, which can limit operational range and flexibility. PPK, by contrast, allows for more extensive operational conditions as the correction process does not need to happen in real time.

- **Accuracy**: Both systems provide high accuracy, but the choice between RTK and PPK may depend on the project's specific needs. RTK can deliver immediate results, which is crucial for projects requiring real-time decision-making. PPK, while delayed, provides an equally high level of accuracy and may be preferable in settings where real-time transmission is less reliable.

- **Cost and Complexity**: RTK setups can be more complex and costly as they require robust communication systems to maintain real-time data links. PPK systems might be more straightforward and cheaper to operate since the data link is not a factor during data collection.

Practical Applications of PPK

PPK technology is particularly advantageous in several critical applications:

- **Remote and Extensive Land Surveying**: In vast or inaccessible areas, such as mountain ranges or dense forests, maintaining a real-time connection between the drone and a base station can be impractical. PPK allows for accurate surveys without the need for a real-time link. This makes it ideal for these environments.

- **Cultural Heritage and Archaeological Documentation**: PPK creates detailed and precise maps of archaeological sites where immediate data processing is not required, but accuracy is paramount. These maps are crucial for preservation, detailed study, and virtually archiving heritage sites.

- **Environmental Monitoring**: PPK aids in environmental conservation efforts, such as tracking coastline changes or glacial retreat over time. Its high accuracy is beneficial for monitoring subtle changes in natural features. This contributes to environmental assessments and climate change studies.

- **Construction and Infrastructure Monitoring**: PPK is utilized to monitor large-scale infrastructure projects. It enables precise and regular updates on construction progress or infrastructure integrity without needing real-time corrections. This facilitates long-term project management and planning.

Post-processing kinematic (PPK) technology provides a robust alternative to RTK. It offers distinct advantages in terms of flexibility, reliability, and operational convenience. As drone technology continues to evolve, the applications of PPK expand, solidifying its role in precision mapping across various sectors. Understanding a project's specific requirements—including the operational environment and data accuracy needs—is crucial in deciding whether PPK or RTK is more suitable. This ensures that the potential of modern GPS technology is fully realized in drone mapping.

Integration of RTK and PPK in Drone Surveys

The integration of Real-Time Kinematic (RTK) and Post-Processing Kinematic (PPK) technologies in drone surveys represents a significant advancement in the field of precision mapping. These technologies significantly enhance the accuracy of geographic data collection. They make them indispensable tools for a wide range of applications. This section examines the synergies between RTK and PPK, outlines the criteria for selecting between them, and presents case studies to illustrate their practical benefits in real-world projects.

Synergies and Selection Criteria

RTK and PPK both enhance GPS accuracy by using a base station that provides corrections. The choice between RTK and PPK often depends on specific project requirements, including the environmental conditions, required data accuracy, and logistical considerations.

- **Real-Time Data Needs**: RTK is ideal for projects requiring immediate data correction and real-time decision-making. It is particularly beneficial in environments where project adjustments must be made on the fly, such as in dynamic construction sites or during emergency response operations.

- **Communication Reliability**: RTK requires a stable communication link between the drone and the base station to receive real-time corrections. If the project area has unreliable communication infrastructure or is susceptible to signal interference (e.g., in urban canyons or densely forested areas), PPK might be a better choice since it does not rely on real-time data transmission.

- **Geographic Extent and Accessibility**: For extensive or remote survey areas, maintaining a real-time correction link (necessary for RTK) can be challenging and resource-intensive. PPK, which processes data post-flight, is more suited to these conditions as it allows for data collection without needing continuous connectivity.

- **Budget and Resource Availability**: Implementing RTK can be more expensive and resource-heavy due to the need for real-time data transmission capabilities. In contrast, PPK can be more cost-effective and simpler to deploy, particularly in resource-limited settings.

In some high-stakes projects, operators combine RTK for real-time navigation with PPK for post-flight refinement, leveraging both immediate feedback and robust post-processing to achieve sub-centimeter precision.

Case Studies

The practical applications of integrating RTK and PPK in drone surveys can be best illustrated through real-world case studies. These demonstrate how these technologies have been successfully employed to achieve high-accuracy results across various sectors.

- **Case Study 1: Infrastructure Inspection**: In a recent project involving the inspection of a large bridge, RTK technology was employed to facilitate immediate assessments of the structure's condition. The real-time data provided by RTK allowed engineers to make on-the-spot evaluations and decisions regarding maintenance needs. This ensured the safety and integrity of the bridge. The ability of RTK to provide instant, precise positioning data was crucial in the rapid identification of potential structural issues. This prevented long-term damage and costly repairs.

- **Case Study 2: Environmental Conservation**: A conservation project aimed at monitoring wetland degradation utilized PPK technology due to the extensive area and the remote location of the wetlands. The project involved mapping changes in vegetation and water levels over time. PPK was selected over RTK due to its suitability for extensive, inaccessible areas where real-time data transmission was not feasible. The post-processed data provided accurate, detailed maps that were critical for assessing the health of the wetlands and planning conservation efforts effectively.

- **Case Study 3: Agricultural Land Management**: In an agricultural project designed to optimize irrigation systems across a vast farm, PPK was integrated to assess topography and optimize water distribution. The high accuracy of the PPK-processed data ensured precise elevation mapping. This was critical for designing an efficient irrigation system that conserved water and reduced costs. The ability to process the data post-flight also allowed for more thorough analysis and adjustments to the system design as needed.

Integrating RTK and PPK technologies into drone surveys has profoundly impacted the precision and utility of geographic data collection. By understanding the specific needs of a project—including the environment, data accuracy requirements, and logistical challenges—professionals can select the most appropriate technology. RTK offers real-time precision, while PPK provides flexibility in data processing. These technologies not only enhance the accuracy of surveys but also expand the potential applications of drone technology across industries. As GNSS technology advances, multi-frequency receivers (e.g., L1/L5 bands) and improved satellite constellations are enhancing RTK and PPK accuracy, promising even greater reliability for drone mapping in the coming years. They pave the way for innovative solutions to complex geographical challenges.

Ground Control Points (GCPs)

To better understand PPK, we must understand ground control points. Ground Control Points (GCPs) are specific points on the ground with known geographic coordinates. They are set using a base station or a base station/rover combination. They are used in aerial mapping to enhance the accuracy and quality of the spatial data collected. These points serve as reference markers in the landscape. They help calibrate and correct aerial images captured by drones, ensuring that the resultant maps and models are precise and reliable.

What Are Ground Control Points?

GCPs are typically marked physically on the ground using high-visibility targets. These range from painted X marks to specialized boards with distinct patterns. The exact coordinates of each GCP are determined by placing the base station on top of an identifiable point and establishing a known location. Once established, that position is recorded and becomes a ground control point. Points can be fixed using a CORS network or GNSS satellite if CORS is unavailable. These coordinates are then used as benchmarks to align and adjust the geospatial data collected during drone flights.

How Are GCPs Used?

In drone mapping, GCPs are essential for achieving high levels of accuracy in photogrammetry projects. After a drone captures aerial photographs, the images are processed using photogrammetry software. During this processing, the software uses the coordinates of the GCPs to correct any geometric distortions in the images. These distortions can be caused by the drone's movement, camera tilt, perspective error, and other factors affecting the accuracy of the measurement.

The use of GCPs is crucial in applications where precise data is necessary, such as construction planning, precision agriculture, or any legal matters where land boundaries and features must be accurately represented. They ensure that the scale of aerial photographs is correct and consistent across the entire mapped area. This is vital for subsequent analysis and decision-making.

How Are GCPs Set?

Setting GCPs is a methodical process that involves several steps:

- **Planning**: Determine how many GCPs are needed and their optimal locations across the survey area to maximize coverage and effectiveness. Typically, it is best to set 1 GCP near each boundary corner, one at the perceived high point, and one at the perceived low point of the property to be mapped. Unless the area is very small, the rule of thumb is that at least 6 GCPs should be deployed. If the land to be surveyed is an abnormal shape, the number should be adjusted accordingly. A common strategy is to distribute GCPs in a grid or zigzag pattern across the site, ensuring even coverage and overlap with flight paths, as visualized in tools like DJI Terra or Pix4D.

- **Placement**: GCPs are placed physically at the chosen locations. This placement must be visible from the air to be identifiable in the drone's imagery. This is the point where a base station and rover setup would be employed. A base station connected to a CORS network or via satellite is used to establish the first known point. A rover is used to measure each GCP point location in relation to the base station.

- **Surveying**: The precise coordinates of each GCP are measured using GNSS technology. For high accuracy, a dual-frequency GPS receiver is often used to mitigate errors caused by the Earth's atmosphere.

- **Mapping**: Once the GCPs are set and their coordinates logged, drone flights can be conducted to collect aerial data. This will later be aligned using the GCPs during the image processing phase.

GPS-Equipped GCPs

Advantages of Propeller AeroPoints Over Standard GCPs

Propeller AeroPoints are a modern innovation designed to enhance the efficiency and accuracy of GCPs in drone operations. Each AeroPoint is an intelligent ground control point that integrates GNSS technology. This makes it significantly more advantageous than standard GCPs. It is, in effect, its own base station. By being equipped with a highly accurate GPS, the AeroPoint can reduce worksite workload by half or more.

- **Ease of Use**: AeroPoints are designed for simplicity and ease of use. Unlike traditional GCPs, which require manual setting and logging, AeroPoints automatically record their exact coordinates when activated. This plug-and-play functionality eliminates the need for specialized surveying knowledge. It makes them accessible to a broader range of users.

- **Accuracy**: Propeller AeroPoints are equipped with high-precision GPS receivers that calibrate in real time to the nearest satellite data. They provide accuracy up to the centimeter level. This is particularly beneficial in projects requiring the highest level of detail and precision. It ensures that any data captured is immediately verifiable and reliable.

- **Time Efficiency**: Using AeroPoints can significantly reduce the time spent on setting and logging GCPs. Traditional methods can be labor-intensive and time-consuming, particularly over large areas or rugged terrain. AeroPoints simplify the process: just place them on the ground, press the button, and they do all the work. They sync data directly to cloud-based platforms for immediate use.

- **Durability and Reusability**: AeroPoints are rugged and weather-resistant. They are designed for use in a variety of environmental conditions. Their durability ensures long-term reusability. This provides cost efficiencies over projects that might otherwise require single-use or less durable markers.

- **Integration**: AeroPoints seamlessly integrate with the Propeller Platform, a comprehensive cloud-based data processing service. This integration allows for immediate processing and visualization of captured data. It facilitates quicker decision-making and enhanced project workflows. Once the geospatial data is processed by Propeller, the data file can then be uploaded to any GCP-matching software. This allows for easy syncing of GCPs during photogrammetry processing.

While traditional GCPs have been invaluable in ensuring the accuracy of aerial mapping, innovations like Propeller AeroPoints offer significant advancements in usability, precision, and efficiency. By reducing the manual effort and expertise required in traditional surveying, AeroPoints save time and improve the reliability and accessibility of drones for high-accuracy mapping projects.

Editor's note:

As we explore the software aspects of drone operations, this text won't provide in-depth tutorials on using various software programs. Instead, we'll focus on outlining their functions, capabilities, and limitations. Many of these software packages come with extensive manuals—often exceeding 100 pages—dedicated solely to explaining their detailed operation. After choosing one or more of these applications, it will be up to you, the reader, to dive into the specifics and master their use.

Chapter 6: Flight Automation Software

In the rapidly evolving world of drone technology, the ability to automate flight operations significantly enhances the efficiency, accuracy, and possibilities of aerial survey, mapping, photogrammetry, and ortho generation. This chapter delves into the diverse realm of flight automation software, exploring cutting-edge platforms that are shaping the future of unmanned flight and aerial data collection. From basic flight planning to advanced autonomous operations, these software solutions offer a range of functionalities tailored to diverse industry needs. These include surveying, façade mapping, severe terrain mapping, construction monitoring, tower inspection, agriculture data collection, and environmental monitoring, all of which can fall into the realm of mapping.

An important aspect to consider when choosing flight automation software is that there is virtually never an apples-to-apples comparison. While most applications have thoroughly overlapping abilities, each typically offers options that differentiate it in a significant way. For instance, when performing a façade mission (3D mapping of structures), UgCS has a complete toolset, making it one of the most capable platforms on the market. However, in other aspects, such as grid mapping, it can be cumbersome and time-consuming to configure for such a simple mission. By comparison, DJI Pilot 2 has an easy-to-use configuration tool for basic ortho mapping but requires using FlightHub 2 on a computer for façade (referred to as "Geometric Route" planning) configuration and flight plan modification. This can be a cumbersome issue when in the field. For this reason, the products discussed in this chapter were chosen; each representing either a drone manufacturer or third-party software in common usage, with distinct features and functionalities in mind.

This chapter provides an in-depth review of several leading flight automation systems, including DJI Pilot 2, Autel Explorer V3, DroneDeploy Flight, Pix4Dcapture Pro, Maps Pilot Pro, DroneLink Growth, and UgCS. Each software brings unique strengths to drone operations, catering to different drone models and specific user requirements. DJI Pilot 2, known for its seamless integration with DJI's drones, offers robust control and mission planning capabilities. In contrast, Autel Explorer V3 serves Autel Robotics users with intuitive interfaces and versatile flight-planning tools. We also explore DroneDeploy Flight and Pix4Dcapture Pro, which are favored by many for their powerful data capture and seamless integration with integrated processing capabilities, making them ideal for detailed geographic information system (GIS) mapping and photogrammetry. Maps Pilot Pro and DroneLink Growth offer many of these capabilities at a lower cost, providing enhanced customization and optimization features that cater to specialized mapping needs for companies on a strict budget. Lastly, UgCS stands out for its compatibility with various drone brands and its ability to manage complex multi-drone missions. It provides a comprehensive solution for large-scale projects and offers near-complete integration with Autel Explorer V3 and DJI Pilot 2, as well as broad compatibility with many drone systems.

Through this exploration, Chapter 6 highlights each flight automation software's technical specifications and operational advantages and discusses their practical applications in real-world scenarios. This discussion aims to equip readers with the knowledge to select the right software that aligns with their operational goals and fully leverages their drone system's potential.

Because flight planning software is so diverse, we will create a list of what are widely considered baseline requirements for legitimate mapping software. We will include this list with each software we examine, noting whether the feature is present and addressing any particular concerns with it.

List of Features and Their Descriptions

When considering flight automation software for mapping drones, several basic features and options are essential for ensuring efficient data collection and optimal mapping outcomes. Here is a comprehensive list of these fundamental capabilities:

- **Integrated Flight Planning**: The ability to customize flight plans according to specific project needs and conditions is absolutely critical in drone mapping operations. Reducing the pilot workload is important while ensuring the data collected meets the required specifications. In most instances, hand-flying is not feasible in commercial mapping operations. This is the most basic requirement for a drone to be considered an enterprise-level mapping or inspection drone.

- **Area Mapping**: Essential for creating detailed maps of specific geographical areas. This feature enables the drone to automatically fly over a defined area and capture images at specified intervals, ensuring complete coverage.

- **Grid Pattern Mapping**: A systematic flight pattern where the drone covers the area in a grid-like fashion. It ensures overlapping coverage crucial for photogrammetry when creating accurate 3D models. This method maximizes area efficiency and data consistency.

- **Waypoint Mapping**: Allows operators to define specific coordinates the drone will navigate to during its flight. This is crucial for missions requiring targeted data collection over particular locations or features within the mapping area.

- **Linear Route Mapping**: Ideal for infrastructure projects such as roads, pipelines, and power lines. This feature enables the drone to follow a linear path while capturing continuous imagery, essential for monitoring and maintenance assessments.

- **Facade Mapping**: Important for structural inspections and architectural surveys. Facade mapping allows the drone to capture detailed images of vertical structures like building faces. This requires precise control over the drone's flight patterns to maintain a consistent distance and angle from the facade.

- **Elevation Hold and Terrain Following**: To effectively handle varied topography, advanced flight automation software should maintain a consistent altitude relative to the ground level. This ensures uniform image quality and accuracy, especially in undulating terrain.

- **Real-Time Monitoring and Adjustment**: Operators should be able to monitor the drone's flight in real time and make necessary adjustments to flight paths, camera settings, or the area of interest. This enhances the flexibility and responsiveness of mapping missions.

- **Airspace Clearances**: In many areas of the U.S., automated approval is available using the FAA's Low Altitude Authorization and Notification Capability (LAANC) system. This provides near real-time approval for flights in controlled airspace. If LAANC is unavailable, you may be required to file an airspace authorization waiver request through the FAA drone website. For other countries, consult the local aviation authority's process.

- **Data Integration and Export Options**: The software should offer seamless integration with various GIS platforms and provide options to export data in compatible formats for further analysis and processing.

- **Safety Features**: Includes geofencing to prevent the drone from entering restricted airspace, automatic return-to-home on loss of signal, and collision avoidance systems to ensure safe operations in complex environments.

- **Obstacle Avoidance Integration**: Integrated within the drone itself, its sensors provide obstacle avoidance when the drone is so equipped.

- **Automatic Takeoff and Landing**: For ease of operation and safety, the ability to automatically take off, execute the mapping mission, and land without manual intervention is extremely useful.

These features represent the foundational tools necessary for effective drone mapping and are indicative of robust flight automation software. Ensuring that your chosen software includes these capabilities will significantly enhance the productivity and accuracy of your aerial mapping projects.

DJI Pilot 2

Advanced flight planning is a cornerstone of professional drone operation. It allows pilots to execute precise and repeatable missions critical for various applications like mapping, surveying, and inspection. DJI Pilot 2 is the proprietary flight automation software for DJI enterprise-level drones. It offers a robust suite of tools designed to facilitate meticulous pre-mission planning, automated flight control, and mission customization. Here is a detailed look at these capabilities, aligned with the baseline requirements:

- **Integrated Flight Planning**: Present. While integrated flight planning is a prerequisite for enterprise drones, DJI Pilot 2 is actively improving its flight planning software to achieve next-level performance. By integrating many features typically found in software costing several thousand dollars or requiring expensive monthly license fees, DJI proves that not only are their drones some of the most capable on the planet, but their software also rivals the best in the business.

- **Area Mapping**: Present. For tasks such as survey, orthomosaic generation, agricultural monitoring, or construction site mapping, DJI Pilot 2 allows users to define a specific area to map. The software automatically calculates the optimal flight path, ensuring complete coverage of the area. It allows the user to choose GSD or altitude based on client requirements, set the percentage of overlap, and adjust the speed of the flight. These options enable optimization of images for desired photogrammetry results. The drone executes the flight autonomously, allowing the pilot to monitor the mission and prepare for unforeseen circumstances.

- **Grid Pattern Mapping**: Present. For tasks requiring advanced image capture for processing into highly accurate 3D orthos, DJI Pilot 2 uses what it refers to as "Oblique." Two types of oblique settings are available: standard "Oblique" and "Smart Oblique." In standard Oblique mode, instead of flying a typical grid pattern, the drone will fly five different data capture flights in varying directions as part of the same mission. In Smart Oblique mode, the drone flies a standard area mapping mission, but the gimbal pans from side to side and back and forth while the drone flies at a slower speed, capturing data from multiple angles. As with standard Area Mapping, all parameters for the maps can be programmed by the pilot.

- **Waypoint Mapping**: Present. One of the most powerful features of DJI Pilot 2 is its ability to set up waypoint missions. Users can program the drone to fly along a predetermined path, which can include multiple waypoints, each with specific actions and camera settings. For instance, at each waypoint, the pilot can define the drone's altitude, speed, orientation, and whether it should pause to capture photos or videos. This is especially useful for aerial surveying and inspections where consistency and coverage are paramount.

- **Linear Route Mapping**: Present. Ideal for infrastructure projects such as roads, pipelines, and power lines, this feature enables the drone to follow a linear path while capturing continuous imagery. This is essential for monitoring and maintenance assessments.

- **Facade Mapping**: Present (via Geometric Route and Slope Route). The Geometric Route, accessible through FlightHub 2, is a recent addition to DJI Pilot 2. It introduces advanced geometric planning tools, ideal for complex architectural and engineering projects. It allows programming of flights based on geometric patterns around structures, facilitating comprehensive data collection for detailed inspections and 3D modeling. Similarly, Slope Route planning, also via FlightHub 2, optimizes flights along varied terrain elevations. It is instrumental in areas like hillside farming or mining where elevation changes are significant. This feature enables drones to maintain a consistent height relative to the slope, enhancing data accuracy and safety in challenging terrains like cliff faces. Note: Accessing these via FlightHub 2 can be cumbersome in the field without a computer.

- **Elevation Hold and Terrain Following**: Present. The application can automatically adjust the drone's altitude to follow the terrain using either an established geoid or imported DTMs. This feature is particularly useful in shifting landscapes where maintaining a consistent altitude above ground level is necessary for accurate data collection, such as in survey or volumetric mapping.

- **Real-Time Monitoring and Adjustment**: Present. DJI Pilot 2 allows users to receive and analyze data in real time, including live video feeds, telemetry data, and sensor outputs. This is essential for making immediate decisions during flight operations, particularly in critical missions like search and rescue or emergency response.

- **Airspace Clearances**: Present (via LAANC integration). DJI Pilot 2 integrates with the FAA's LAANC system for near real-time airspace approvals in controlled U.S. airspace, streamlining compliance.

- **Data Integration and Export Options**: Present. The software supports integration with FlightHub 2 and offers export options for various GIS platforms, enhancing workflow flexibility.

- **Safety Features**: Present. Includes geofencing, automatic return-to-home on loss of signal or low battery, and robust safety protocols to ensure compliance and secure operations.

- **Obstacle Avoidance Integration**: Present. The application integrates DJI's obstacle avoidance technologies, which vary depending on the drone model (e.g., visual, laser, or radar-based systems), enhancing safety in complex environments.

- **Automatic Takeoff and Landing**: Present. DJI Pilot 2 supports fully automated takeoff and landing, simplifying operations and reducing manual intervention risks.

Additional Unique Features:

1. **DJI FlightHub 2 Integration**: Enhances desktop mission planning, multi-drone coordination, and live streaming capabilities.

2. **Mission Sharing and Collaboration**: Through FlightHub 2, DJI Pilot 2 allows sharing mission plans among team members, ensuring coordination in large-scale operations.

3. **Customization through SDK**: DJI offers a Software Development Kit (SDK) that integrates with DJI Pilot 2. This allows for custom flight planning modules tailored to specific industrial applications, further extending the software's functionality.

4. **Enhanced Compatibility with DJI Hardware**: Finely tuned to work seamlessly with DJI's enterprise drone hardware, optimizing control and performance.

5. **Integrated Thermal Imaging**: Offers seamless switching between RGB and thermal views or simultaneous side-by-side displays for drones with thermal cameras.

6. **Localized Data Mode**: Ensures data privacy by preventing transmission to DJI or third parties, ideal for sensitive operations.

7. **Extensive Accessory and Sensor Support**: Supports a wide range of DJI and third-party accessories (e.g., thermal, LiDAR, loudspeakers), enhancing mission versatility.

DJI Pilot 2 is a robust flight planning and management application designed specifically for DJI's enterprise drone models, such as the Matrice and Mavic series. It caters to professional users in sectors like infrastructure, energy, and agriculture. With its comprehensive suite of advanced features, seamless hardware integration, and enhanced safety capabilities, it stands as a top choice for professional drone operations across various industries.

DJI FlightHub 2

While DJI FlightHub 2 is flight planning software designed to integrate directly with most DJI drones, it is much more. It is a drone fleet management system offering live video feeds from drones to any internet connection, real-time telemetry tracking, and the ability to remotely modify and upload flight plans to drones in the field in seconds. It even includes live mapping—an ability to upload images from the ground control station (remote) to the cloud and generate 2D orthos on the fly. While this isn't practical for daily engineering operations, it is highly effective in search and rescue. Although other fleet management systems exist, FlightHub 2 exemplifies the future of drone technology. When integrated with tools like the DJI Dock and Dock 2, FlightHub 2 enables completely autonomous remote drone operations. This literally allows an operations manager in one part of the world to launch and control a flight halfway around the globe with no local interaction.

While some versions have a cost associated with them, the free version is amazingly functional (free as of May 2024). It includes live telemetry, desktop flight planning, and instant modification and upload of mission plans to drones in the field.

Feature Evaluation:

- **Integrated Flight Planning**: Present. Offers advanced desktop-based planning, complementing DJI Pilot 2 for complex missions.

- **Area Mapping**: Present. Supports area mapping with real-time 2D ortho generation, though optimized for situational awareness rather than detailed engineering outputs.

- **Grid Pattern Mapping**: Present. Available via integration with DJI Pilot 2, though not a standalone feature in FlightHub 2.

- **Waypoint Mapping**: Present. Supports waypoint planning through desktop tools, uploadable to drones in the field.

- **Linear Route Mapping**: Present. Facilitates linear missions, especially for infrastructure, via desktop planning.

- **Facade Mapping**: Present (via Geometric/Slope Routes). Accessible through FlightHub 2's desktop interface, though field adjustments require a computer, which can be inconvenient.

- **Elevation Hold and Terrain Following**: Present. Inherits DJI Pilot 2's terrain-following capabilities when missions are executed.

- **Real-Time Monitoring and Adjustment**: Present. Exceptional real-time telemetry, live streaming, and remote flight plan adjustments set it apart.

- **Airspace Clearances**: Present. Integrates LAANC for U.S. airspace approvals, streamlining regulatory compliance.

- **Data Integration and Export Options**: Present. Cloud-based storage and export options enhance data sharing and integration with GIS platforms.

- **Safety Features**: Present. Includes geofencing and return-to-home features, bolstered by real-time oversight.

- **Obstacle Avoidance Integration**: Present. Relies on drone hardware capabilities, managed via FlightHub 2's interface.

- **Automatic Takeoff and Landing**: Present. Fully supported, especially with DJI Dock systems for autonomous operations.

Unique Features:

1. **Real-Time Operations Management**: Live streaming and flight data monitoring enable instant decision-making.

2. **Fleet Management**: Tracks drones, pilots, and maintenance logs, optimizing large-scale operations.

3. **Cloud-Based Collaboration**: Secure data storage and team collaboration tools enhance workflow efficiency.

4. **Scalability**: Scales from a few drones to hundreds, ideal for enterprise use.

Limitations: Limited to Mavic 3 Enterprise and Matrix 30 series for full mapping compatibility (workaround: build maps for compatible drones and adjust for others like the M350 RTK, double-checking settings). Requires reliable internet, which may limit use in remote areas.

DJI FlightHub 2 is a powerful drone operations management platform that significantly enhances enterprises' ability to use and manage their drone fleets. Its extensive functionality makes it ideal for sectors requiring robust, real-time, and collaborative drone operations. Despite the cost and need for consistent internet connectivity, its benefits in operational efficiency, safety, and data management make it a valuable investment for organizations leveraging drone technology at scale.

Autel Explorer V3

Many experienced drone pilots may not realize that Autel offers an extensive mapping platform for its enterprise drones. Autel Explorer V3 is the proprietary flight management software developed for Autel's enterprise-level drones. It is designed to support complex, high-precision tasks across industries such as surveying, agriculture, public safety, and infrastructure inspection. It integrates a suite of sophisticated tools for pre-mission planning, automated flight control, and tailored mission customization. Here's a detailed exploration of these capabilities:

- **Integrated Mission Planning**: Present. Autel Explorer V3 excels in creating customizable flight plans. The software allows users to design specific flight paths and adjust parameters to meet the unique requirements of different missions. This flexibility is crucial for tailored applications in various industries, from agriculture to construction.

- **Area Mapping**: Present. The software supports extensive area mapping, enabling drones to cover large regions efficiently. This detailed map generation is ideal for applications like surveying and environmental monitoring, where accurate and expansive data collection is necessary.

- **Grid Pattern Mapping**: Present. Autel Explorer V3's grid pattern mapping feature is highly effective for systematic data collection. This is particularly useful for agricultural applications, where consistent and evenly spaced data points are essential for crop analysis and management.

- **Waypoint Mapping**: Present. The waypoint mapping capability allows precise navigation and data collection at specified locations. This feature is invaluable for repeatable missions, ensuring the drone follows the same path each time, which is essential for longitudinal studies and monitoring projects.

- **Linear Route Mapping**: Present. The software supports linear route mapping, making it suitable for inspections of linear structures like pipelines, roads, and power lines. This ensures thorough and consistent data collection along the entire route.

- **Facade Mapping**: Present. Autel Explorer V3 includes facade mapping capabilities, allowing drones to capture detailed images of vertical structures. This is particularly beneficial for building inspections and architectural assessments, providing high-resolution data of building exteriors.

- **Elevation Hold and Terrain Following**: Present. The elevation hold and terrain following features ensure drones maintain a consistent altitude relative to the ground. This capability is crucial for missions over uneven terrain, providing accurate data and preventing collisions.

- **Real-Time Monitoring and Adjustment**: Present. The real-time monitoring and adjustment features allow operators to oversee the drone's flight and make necessary adjustments on the fly. This ensures missions can adapt to changing conditions or requirements, enhancing flexibility and efficiency.

- **Airspace Clearances**: Present (manual process). While it lacks direct LAANC integration, operators can manage airspace approvals manually via local authorities, supported by the software's planning tools.

- **Data Integration and Export Options**: Present. Autel Explorer V3 offers robust data integration and export options, supporting various formats for easy data sharing and analysis. This interoperability is essential for integrating drone data into broader workflows and applications.

- **Safety Features**: Present. The software incorporates multiple safety features, including geofencing, return-to-home, and low battery warnings. These enhance operational safety, ensuring the drone can avoid restricted areas and return safely in case of emergencies.

- **Obstacle Avoidance Integration**: Present. Autel Explorer V3 integrates obstacle avoidance technology, reducing the risk of collisions during flights. This feature is critical for maintaining safety in complex environments, leveraging the drone's sensor suite.

- **Automatic Takeoff and Landing**: Present. The automatic takeoff and landing feature simplifies drone operation, making it accessible even for less experienced users. This automation enhances the user experience and reduces the likelihood of errors during these critical phases.

Unique Features:

1. **Integrated Hardware and Software Solution**: Designed to work seamlessly with Autel Robotics' drones, optimizing both hardware and software capabilities for smoother performance.

2. **Advanced Obstacle Avoidance**: Features multiple sensors (front, rear, side, downward-facing) for real-time obstacle detection, on par with top-tier systems.

3. **Thermal Imaging Integration**: Supports thermal imaging with swappable payloads on some models (e.g., Evo II), ideal for inspections and search and rescue.

4. **6K Video Resolution**: Captures up to 6K native video, offering superior image quality for professional-grade productions.

5. **Dynamic Track 2.1**: Advanced tracking technology for automatically following moving subjects with high precision, beneficial for dynamic scenes.

6. **Precision Flight Modes**: Includes Waypoint Flight, Orbit, and Precision Flight for enhanced control over complex missions.

7. **Dual Control Mode**: Allows two pilots to share control over flight and camera operations, useful for cinematography and search and rescue.

8. **No Geofencing Restrictions**: Lacks built-in geofencing, offering flexibility but requiring operator diligence for regulatory compliance.

Autel Explorer V3 offers specialized capabilities catering to professional drone users needing advanced imaging, tracking, and integration. Its customizable flight plans, robust mapping features, real-time monitoring, and safety enhancements make it a valuable tool for surveying, inspections, and environmental monitoring. With a cost of entry below $7,000 (May 2024), it provides a compelling option for budget-conscious professionals.

DroneDeploy Flight

DroneDeploy Flight is a robust drone software solution designed for various mapping and surveying applications. Here's an evaluation based on the baseline features:

- **Integrated Mission Planning**: Present. DroneDeploy offers highly customizable flight plans, essential for diverse operational needs ranging from agriculture to construction. Users can modify flight paths, altitudes, and speeds directly within the app, making it highly adaptable to various project requirements.

- **Area Mapping**: Present. DroneDeploy excels in area mapping with automated flight planning tools that enable users to capture high-resolution aerial images. These are particularly useful for topographic surveys, agricultural assessments, and construction site monitoring.

- **Grid Pattern Mapping**: Present. The platform supports grid pattern mapping, critical for ensuring complete area coverage. This feature is vital for agricultural monitoring, where consistent grid patterns help capture uniform data across large fields.

- **Waypoint Mapping**: Present. Waypoint mapping in DroneDeploy allows for precise path planning by setting multiple waypoints along the desired flight route. This aids in detailed site inspections and specific point-of-interest monitoring.

- **Linear Route Mapping**: Present. DroneDeploy's linear route mapping is advantageous for infrastructure inspections, such as pipelines and roads. The software simplifies planning linear missions, aiding regular maintenance checks and quick assessments.

- **Facade Mapping**: Present. For vertical structures like building facades, DroneDeploy offers specialized flight plans that capture detailed images from various angles. This is crucial for construction monitoring, architectural assessments, and historical preservation efforts.

- **Elevation Hold and Terrain Following**: Present. The elevation hold and terrain following capabilities ensure drones maintain a consistent altitude relative to the ground level. This is crucial for acquiring stable and uniform imagery over uneven terrain.

- **Real-Time Monitoring and Adjustment**: Present. DroneDeploy provides real-time flight monitoring, allowing operators to make on-the-fly adjustments to the flight path. This responsiveness is critical for adapting to changing conditions or focusing on areas needing additional attention.

- **Airspace Clearances**: Present. Integrates with LAANC for automated airspace approvals in the U.S., enhancing compliance efficiency.

- **Data Integration and Export Options**: Present. The software integrates seamlessly with various GIS and CAD software, providing robust data export options essential for professional analysis and reporting.

- **Safety Features**: Present. Includes geofencing to prevent drones from entering restricted airspace and automatic weather checks before flights to ensure safe operating conditions.

- **Obstacle Avoidance Integration**: Present (hardware-dependent). DroneDeploy supports data from sensors for obstacle avoidance, though effectiveness depends on the drone's hardware capabilities.

- **Automatic Takeoff and Landing**: Present. Fully supports automated takeoff and landing, enhancing ease of use and safety.

Unique Features:

1. **Ease of Use and Accessibility**: Known for its user-friendly interface, reducing the learning curve for beginners and experienced operators alike.

2. **Comprehensive Mobile Integration**: Offers full functionality via a mobile app, allowing operation, real-time feeds, and data analysis from smartphones or tablets.

3. **Advanced Data Processing**: Rapidly processes captured data into actionable insights, generating reports, volume calculations, and 3D models.

4. **High-Quality Mapping**: Uses sophisticated photogrammetry for high-resolution maps and models, ideal for precise measurements.

5. **Seamless Third-Party Integration**: Robust APIs and a marketplace connect with tools like SAP and Autodesk, enhancing workflow.

6. **Live Map Technology**: Real-time mapping without internet connectivity, useful for immediate ground truthing.

7. **Versatile Flight Planning Options**: Includes polygon, corridor, and vertical mapping modes for project-specific customization.

8. **Strong Community and Support**: Offers extensive resources, training, and a vibrant user community for collaboration and troubleshooting.

DroneDeploy stands out for its ease of use, comprehensive data capabilities, and real-time functionalities. Its strong third-party integrations and high-quality mapping position it as a top choice for professional users in agriculture, construction, and mining. Performance in obstacle avoidance depends on drone hardware, which users should consider when evaluating its effectiveness.

Pix4Dcapture Pro

Pix4Dcapture Pro is a drone flight planning software designed for photogrammetry missions. It is used extensively in agriculture, construction, and surveying industries. It is particularly noted for integrating with Pix4D's photogrammetry suite, enabling a seamless transition from flight planning to data processing. Here's a detailed evaluation:

- **Integrated Mission Planning**: Present. Pix4Dcapture Pro offers robust mission planning tools that integrate with Pix4D's processing software. This ensures that data captured during flights is optimized for subsequent analysis and modeling, streamlining the process for professional users.

- **Area Mapping**: Present. The software excels in area mapping with automated flight plans that adjust to specific project requirements. It allows users to define the area to be mapped with simple touch interactions, automatically calculating the optimal flight path for efficient coverage.

- **Grid Pattern Mapping**: Present. Grid pattern mapping is well-executed, providing thorough coverage of the mapped area. This is crucial for applications like agriculture, where consistent data coverage is necessary for analyzing crop health and other variables.

- **Waypoint Mapping**: Present. Waypoint mapping allows users to manually set waypoints for controlled and specific flight paths. This is particularly useful in complex environments or for detailed inspections of specific areas.

- **Linear Route Mapping**: Present. For projects involving infrastructure like roads and pipelines, Pix4Dcapture Pro offers effective linear route mapping. It plans flights along a linear path, capturing detailed, continuous imagery for inspection and monitoring.

- **Facade Mapping**: Present. Facade mapping capabilities enable the capture of detailed vertical structures like building facades. The software provides tools to adjust the drone's flight path and camera angle, ensuring high-quality data collection for structural analysis and modeling.

- **Elevation Hold and Terrain Following**: Present. Supports elevation hold and terrain following, critical for maintaining consistent altitude over uneven terrain. This ensures uniform quality of imagery and data captured.

- **Real-Time Monitoring and Adjustment**: Present. Real-time flight data monitoring and the ability to adjust parameters on the fly are well-supported, enhancing safety and efficiency.

- **Airspace Clearances**: Present (manual process). Lacks direct LAANC integration but supports manual airspace authorization planning.

- **Data Integration and Export Options**: Present. Integrates seamlessly with Pix4D's processing tools and supports various export options for compatibility with GIS and CAD applications.

- **Safety Features**: Present. Includes geofencing, automatic return-to-home on low battery or lost connection, and weather condition checks to prevent accidents and ensure compliance.

- **Obstacle Avoidance Integration**: Present (hardware-dependent). Provides basic support for obstacle avoidance, with effectiveness tied to the drone's sensor capabilities.

- **Automatic Takeoff and Landing**: Present. Fully supports automated takeoff and landing, enhancing operational ease and safety.

Unique Features:

1. **Seamless Photogrammetry Integration**: Designed to complement Pix4Dmapper, optimizing data for 3D reconstruction and analysis with direct cloud upload.

2. **Specialized Mapping Modes**: Offers grid, double grid, circular, and corridor modes tailored to specific survey types for high-quality data capture.

3. **High Precision and Accuracy**: Focuses on maximizing image overlap and resolution for precise photogrammetry results.

4. **Advanced Camera Control**: Allows detailed adjustments to overlap, sidelap, altitude, and angle, ensuring data meets project needs.

5. **Offline Mission Planning**: Supports planning without an internet connection, ideal for remote areas.

6. **Multi-Drone Support**: Compatible with a wide range of drones, offering flexibility in hardware choice.

7. **Data Quality Assurance**: Checks camera focus, GPS health, and flight parameters to ensure high-quality data collection.

Pix4Dcapture Pro is a comprehensive flight planning tool for photogrammetry. Its strong mapping capabilities, seamless data integration, and robust safety features make it an excellent choice for professional drone operators. Users must pair it with compatible drones equipped with necessary sensors to fully leverage its capabilities, especially for obstacle avoidance.

Maps Pilot Pro

Maps Pilot Pro, designed for drone flight planning and data capture, serves users requiring basic mapping functionalities. Compared to more advanced systems like Pix4Dcapture Pro or DJI Pilot 2, it offers a streamlined approach focusing on fundamental needs. Here's a detailed evaluation:

- **Integrated Mission Planning**: Present (basic). Provides basic tools suitable for straightforward projects. It lacks the depth and complexity of advanced software, limiting its use in professional settings requiring detailed customization.

- **Area Mapping**: Present (basic). Supports basic area mapping, allowing users to define the mapping area manually. It lacks advanced automation features to optimize flight paths based on area characteristics.

- **Grid Pattern Mapping**: Present (basic). Offers grid pattern mapping for general purposes like agriculture or 3D photogrammetry. The options are less customizable than in sophisticated software, potentially impacting efficiency and coverage.

- **Waypoint Mapping**: Not Present. Without waypoint mapping, users have limited control over specific flight paths, a drawback for detailed inspections or targeted surveys.

- **Linear Route Mapping**: Not Present. Lacks linear route mapping, limiting its utility for infrastructure inspections like roads and pipelines where precise linear data capture is essential.

- **Facade Mapping**: Not Present. Absence of facade mapping makes it less suitable for architectural and construction applications requiring vertical structure data.

- **Elevation Hold and Terrain Following**: Present (basic). Provides basic elevation hold but lacks advanced terrain following. This can affect data uniformity over uneven terrain, compromising accuracy.

- **Real-Time Monitoring and Adjustment**: Present (basic). Offers basic real-time monitoring but lacks in-depth adjustment capabilities, limiting responsiveness in dynamic environments.

- **Airspace Clearances**: Present (manual process). Supports manual airspace planning but lacks automated LAANC integration.

- **Data Integration and Export Options**: Present (limited). Supports standard export options, but integration with GIS or other systems is limited, a drawback for comprehensive workflows.

- **Safety Features**: Present (basic). Includes return-to-home on low battery or lost connection but lacks advanced features like geofencing or weather checks.

- **Obstacle Avoidance Integration**: Not Present. Relies solely on the drone's built-in systems without software-level integration, posing risks in complex scenarios.

- **Automatic Takeoff and Landing**: Present. Supports automated takeoff and landing, simplifying basic operations.

Unique Features:

1. **Simplicity and User-Friendliness**: Designed with an intuitive interface, making it accessible to novices and hobbyists.

2. **Focused Functionality**: Emphasizes essentials for basic mapping, avoiding complexity.

3. **Efficient Area Mapping**: Handles basic area mapping efficiently for small-scale tasks.

4. **Cost-Effectiveness**: More budget-friendly than advanced counterparts, appealing to cost-conscious users.

5. **Direct Data Handling**: Enables direct downloading and handling of data, simplifying retrieval.

6. **Quick Setup for Repeat Missions**: Allows saving flight plans for quick reloading, useful for monitoring changes over time.

Maps Pilot Pro is a straightforward, user-friendly app suited for basic mapping tasks. Its simplicity appeals to hobbyists or beginners, but it lacks advanced features for professional use. Users needing detailed planning, complex integration, or robust safety mechanisms should consider more sophisticated solutions.

DroneLink Growth

DroneLink Growth is a comprehensive drone flight planning and execution software known for its advanced mission planning and data capture capabilities. It is also noted as resource-intensive and occasionally buggy, affecting performance. Here's an evaluation:

- **Integrated Mission Planning**: Present. Provides robust tools for detailed flight operations planning. Its interface supports complex requirements, enabling precise paths, actions, and camera settings. However, this complexity can demand significant hardware resources, potentially slowing performance on less capable devices.

- **Area Mapping**: Present. Excels in area mapping with options to configure detailed grid patterns for efficient coverage. This is essential for agricultural monitoring and environmental mapping, ensuring high-quality data collection.

- **Grid Pattern Mapping**: Present. Supports advanced grid pattern mapping with customizable settings for overlap, altitude, and speed. These are tailored for optimal photogrammetry and 3D modeling, though precision can strain hardware.

- **Waypoint Mapping**: Present. Well-executed waypoint mapping allows defining specific points and detailed paths. This is crucial for inspections and surveillance, offering complete control.

- **Linear Route Mapping**: Present. Supports linear route mapping, ideal for inspecting linear infrastructure. Its ability to plan and execute can be affected by occasional bugs, potentially impacting reliability.

- **Facade Mapping**: Present. Offers specialized tools for facade mapping, capturing detailed vertical structures. High processing demands and potential bugs can affect smooth execution.

- **Elevation Hold and Terrain Following**: Present. Includes features to maintain consistent altitude in varied landscapes, aiding uniform data capture. These require significant processing power.

- **Real-Time Monitoring and Adjustment**: Present. Allows real-time monitoring and adjustments, vital for dynamic conditions. Software issues may hamper responsiveness.

- **Airspace Clearances**: Present (manual process). Supports manual airspace planning but lacks automated LAANC integration.

- **Data Integration and Export Options**: Present. Offers comprehensive integration and export options, facilitating compatibility with other systems.

- **Safety Features**: Present. Includes geofencing, automatic return-to-home, and battery management. Stability issues could affect reliability.

- **Obstacle Avoidance Integration**: Present (variable). Relies on drone capabilities, with integration varying by model; glitches may compromise effectiveness.

- **Automatic Takeoff and Landing**: Present. Fully supports automated takeoff and landing, enhancing ease of use.

Unique Features:

1. **Highly Customizable Missions**: Allows detailed scripting of behaviors, surpassing typical settings in other software.

2. **Cross-Platform Compatibility**: Supports various drone models across manufacturers, offering flexibility.

3. **Complex Flight Path Design**: Executes intricate patterns (e.g., spirals, figures of eight), ideal for cinematography.

4. **Scripting and Automation**: Enables custom scripts for precise control, unique among mapping tools.

5. **Educational and Community Support**: Fosters a community for sharing scripts and tips, enhancing learning.

6. **Resource-Intensive**: Demands high-performance hardware, reflecting its advanced capabilities but posing challenges for some users.

DroneLink Growth is a powerful tool for sophisticated drone operations. Its extensive functionalities suit professional use, but its resource-heavy nature and occasional bugs may impact performance. Users with high-performance hardware and advanced needs will find it highly beneficial, while those requiring reliability might face challenges.

Summary and Conclusion

In the rapidly evolving domain of drone technology, flight automation software stands out as a crucial element. It significantly enhances the efficiency, accuracy, and possibilities of aerial survey, mapping, photogrammetry, and ortho generation. This chapter has provided a basic overview of several flight automation platforms, highlighting their unique features and capabilities.

We've explored leading systems like DJI Pilot 2, Autel Explorer V3, DroneDeploy Flight, Pix4Dcapture Pro, Maps Pilot Pro, DroneLink Growth, and UgCS (not detailed here but noted for its multi-drone and facade mapping strengths). Each offers a distinct set of tools tailored to industries such as surveying, façade mapping, severe terrain mapping, corridor mapping, construction monitoring, tower inspection, agriculture data collection, and environmental monitoring.

No single software solution fits all scenarios. Each platform has strengths and limitations, making it essential to understand your operational requirements before choosing a tool. For instance, UgCS excels in complex multi-drone missions and façade mapping, while DJI Pilot 2 is noted for seamless DJI integration and advanced planning. DroneDeploy and Pix4Dcapture Pro are favored for powerful data capture and integration, ideal for GIS and photogrammetry.

Baseline requirements for effective mapping—integrated flight planning, area and grid mapping, waypoint and linear route mapping, facade mapping, elevation hold, real-time monitoring, and data integration—enhance productivity and accuracy. The ability to adjust these features to specific project needs is crucial for optimal outcomes.

As drone technology advances, flight automation software's role becomes increasingly vital. This chapter equips you with the knowledge to make informed decisions, leveraging these tools to maximize your drone system's potential. By aligning platform nuances with operational goals, you can unlock new levels of efficiency, accuracy, and innovation in aerial data collection.

Chapter 7: Photogrammetry Software

Introduction to Photogrammetry Software

Photogrammetry harnesses photographs to measure and interpret objects and their surroundings, delivering precise spatial insights. In the realm of drone mapping, photogrammetry software transforms overlapping aerial images into georeferenced 2D maps, 3D models, and other critical datasets. Leveraging advanced algorithms and geometric principles, these platforms efficiently produce accurate geospatial outputs, empowering professionals across diverse industries.

Historically rooted in surveying, mapping, and archaeology, photogrammetry has been transformed by drone technology. Equipped with high-resolution cameras, drones now capture expansive datasets swiftly, accessing perspectives once out of reach. This evolution has expanded photogrammetry's utility to agriculture, construction, mining, environmental monitoring, and public safety, broadening its impact on modern workflows.

At its core, photogrammetry software automates the conversion of raw drone imagery into actionable intelligence. Techniques like Structure from Motion (SfM) enable the creation of point clouds, mesh models, and orthomosaics, offering tangible benefits: precise measurements, change detection over time, and informed decision-making. From generating topographic maps to monitoring crop health or guiding construction planning, these tools are indispensable for contemporary spatial analysis.

As the demand for precision mapping intensifies, photogrammetry software continues to evolve, with platforms offering specialized capabilities—from streamlined 2D mapping to intricate 3D reconstructions. These advancements amplify drone potential, pushing the boundaries of aerial data collection and application.

This chapter evaluates leading photogrammetry software platforms, each distinguished by features tailored to specific industry needs. Yet, photogrammetry is not the sole method for point cloud generation; technologies like LiDAR also play a pivotal role. Following our review of photogrammetry tools, we'll explore these complementary approaches and their supporting solutions, providing a comprehensive view of the geospatial landscape.

DJI Terra

DJI Terra is a powerful mapping software that converts drone-captured data into actionable digital assets, serving industries such as agriculture, construction, and surveying. This review examines its advanced photogrammetric and LiDAR processing capabilities, point cloud and GeoTIFF generation, and seamless GIS integration, highlighting its precision and utility for professional geospatial workflows.

Key Features and Capabilities

DJI Terra delivers a comprehensive suite of tools:

- **Photogrammetry**: Processes DJI drone imagery into detailed 3D reconstructions, including point clouds, textured models, and terrain maps, essential for precision modeling.

- **LiDAR Processing**: Integrates LiDAR data to produce accurate elevation models, penetrating vegetation for forestry and environmental analysis.
- **Real-Time Mapping**: Supports on-site, real-time 2D and 3D mapping, critical for rapid-response scenarios like disaster assessment.
- **Mission Planning**: Automates flight paths and parameters, ensuring consistent, high-quality data collection.

Industry-Specific Applications

DJI Terra adapts to diverse professional needs:

- **Construction**: Provides 3D models and terrain analysis for site planning and volume estimation.
- **Agriculture**: Generates orthomosaics and vegetation maps to optimize land use and crop management.
- **Surveying**: Delivers survey-grade data for land management, urban planning, and legal documentation.
- **Environmental Studies**: Leverages LiDAR to map complex terrains and monitor ecological changes.

Rich Array of Deliverables

The platform produces versatile, high-value outputs:

- **Orthomosaic Maps**: Georeferenced 2D maps for spatial analysis and planning.
- **3D Models**: Detailed visualizations for construction oversight and public safety assessments.
- **Point Clouds**: Dense, accurate datasets for elevation and volumetric studies.
- **Terrain Maps**: Precise digital representations for landscape and urban design.

Advantages of Integration and Precision

DJI Terra excels in compatibility and accuracy:

- **GIS Software Compatibility**: Exports LAS/LAZ point clouds and GeoTIFF orthomosaics, integrating effortlessly with ArcGIS and QGIS for advanced geospatial workflows.
- **Survey-Grade Precision**: Ground Control Point (GCP) support and LiDAR processing ensure reliable, high-accuracy outputs for regulatory and legal applications.
- **Efficiency**: Combines real-time mapping with robust post-processing, accelerating project timelines.

Thoughts

DJI Terra stands out as a versatile and precise solution for transforming drone data into actionable insights. Its advanced photogrammetry, LiDAR capabilities, and GIS-ready exports make it a vital tool for surveyors, planners, and drone operators. As aerial technology evolves, DJI Terra is poised to play an increasingly central role in enhancing spatial analysis and operational efficiency across industries.

Autel Mapper

Autel Mapper is a robust drone data processing platform tailored for Autel Robotics' drones, converting raw aerial imagery into precise, actionable geographic outputs. Designed for industries such as surveying, construction, agriculture, and environmental monitoring, it combines powerful photogrammetric tools with seamless GIS integration. This review examines its core functionalities—point cloud generation, GeoTIFF production, and GIS export capabilities—and evaluates their practical applications for professional workflows.

Key Features and Capabilities

Autel Mapper excels in delivering specialized geospatial tools:

- **Photogrammetry and 3D Modeling**: Leveraging advanced algorithms, it transforms high-resolution drone imagery into detailed 3D models and maps. This precision supports applications like urban planning, construction site analysis, and archaeological documentation.
- **Point Cloud Generation**: The platform produces dense, high-fidelity point clouds, enabling the creation of digital elevation models essential for geological assessments, forestry management, and infrastructure design.
- **GeoTIFF Output**: It generates GeoTIFF files embedded with georeferenced data, ensuring spatial accuracy for GIS-based tasks such as environmental impact studies and land-use planning.
- **Automated Flight Planning**: Integrated mission planning tools allow users to define flight paths and parameters, optimizing data consistency across expansive survey areas.

User Experience and Performance

The software balances accessibility with performance:

- **Intuitive Interface**: A streamlined design guides users from mission setup to data analysis, making it approachable for beginners while meeting the demands of seasoned professionals.
- **Processing Efficiency**: Optimized to handle large datasets swiftly, it delivers accurate results without compromising on detail—a critical advantage for time-sensitive projects like disaster response or construction deadlines.

Integration and Professional Outputs

Autel Mapper shines in its compatibility with industry-standard tools:

- **GIS Software Compatibility**: Exports include LAS-format point clouds and GeoTIFF orthomosaics, integrating effortlessly with platforms like ArcGIS and QGIS for advanced spatial analysis.
- **Survey-Grade Precision**: Support for Ground Control Points (GCPs) ensures outputs meet rigorous standards, making it suitable for cadastral mapping, regulatory reporting, and construction oversight.

Thoughts

Autel Mapper stands out as a versatile solution for transforming aerial data into high-value geographic insights. Its blend of point cloud generation, GeoTIFF production, and GIS-ready exports positions it as a reliable asset across diverse sectors. As drone technology advances, Autel Mapper is poised to play an increasingly vital role in enhancing the efficiency and accuracy of geospatial workflows.

DroneDeploy

DroneDeploy is a premier cloud-based platform that transforms drone-captured aerial data into actionable insights, serving industries like construction, agriculture, surveying, and real estate. This review examines its robust capabilities—point cloud and GeoTIFF generation, GIS exports, and a diverse range of outputs including orthomosaics, 3D models, videos, and panoramas—underscoring its adaptability for professional use.

Key Features and Capabilities

DroneDeploy leverages cutting-edge tools for geospatial precision:

- **Point Cloud Generation**: It processes imagery into detailed point clouds, enabling 3D representations for digital elevation models and volumetric analysis in mining and construction.
- **GeoTIFF Output**: The platform produces GeoTIFF files with embedded geographic metadata, ensuring accuracy for GIS-driven spatial analysis.
- **GIS Software Compatibility**: Exports such as LAS/LAZ point clouds and GeoTIFF orthomosaics integrate seamlessly with tools like ArcGIS and QGIS.

Industry-Specific Applications

DroneDeploy tailors its functionality to diverse sectors:

- **Construction Monitoring**: Tracks project progress, measures stockpile volumes, and flags discrepancies, supporting on-time and on-budget delivery.
- **Agriculture**: Generates NDVI and moisture maps to optimize resource use, enhancing crop yields and reducing costs.
- **Surveying**: Streamlines geospatial data collection for land management and conservation projects.
- **Environmental Monitoring**: Monitors ecosystem shifts and assesses post-disaster impacts with precision.

Rich Array of Deliverables

The platform offers a versatile suite of outputs:

- **Orthomosaic Maps**: High-resolution, georeferenced maps for in-depth site analysis.
- **3D Models**: Interactive visualizations ideal for planning and virtual site tours.
- **Videos and Photos**: Professional-grade media for inspections and marketing purposes.
- **Panoramic Views**: 360-degree imagery providing immersive perspectives of surveyed areas.

Advantages of a Web-Based Platform

DroneDeploy's cloud infrastructure enhances its utility:

- **Global Access**: Enables project management from any location, supporting remote workflows.
- **Scalability**: Adapts effortlessly to projects of varying scope, from small plots to vast regions.
- **Collaboration**: Real-time data sharing accelerates decision-making across teams.
- **Continuous Improvement**: Automatic updates ensure access to the latest features.

Thoughts

DroneDeploy stands out as a dynamic solution for converting aerial imagery into high-impact geospatial data. Its comprehensive outputs, cloud-based flexibility, and industry-specific tools make it a cornerstone for professionals. As drone technology evolves, DroneDeploy is well-positioned to drive operational efficiency and innovation across sectors.

Pix4D

Pix4D is a leading photogrammetry suite that transforms aerial imagery into detailed geospatial outputs, catering to surveying, construction, agriculture, and infrastructure management. This review explores its specialized platform variants, GIS integration, industry applications, and comprehensive deliverables, highlighting the advantages of its cloud-based capabilities.

Key Features and Capabilities

Pix4D offers a tailored ecosystem of tools for precision mapping:

- **Platform Variants**:

  - *Pix4Dmapper*: The flagship solution generates maps and 3D models with volume and contour tools, ideal for surveying and construction.
  - *Pix4Dfields*: Focuses on agriculture, delivering crop health analytics for precision farming.
  - *Pix4Dreact*: Designed for rapid 2D mapping in emergency response scenarios like disaster assessment.
  - *Pix4Dinspect*: Automates industrial inspections, such as power line monitoring, for asset management.
  - *Pix4Dsurvey, Pix4Dmatic, Pix4Dcloud*: Support large-scale surveying, LiDAR integration, and cloud-based project oversight.

- **GIS Software Compatibility**: Exports GeoTIFF orthomosaics, LAS point clouds, and OBJ models, integrating seamlessly with ArcGIS and QGIS.

Industry-Specific Applications

Pix4D adapts its functionality to diverse professional needs:

- **Surveying and Construction**: Pix4Dmapper and Pix4Dsurvey deliver topographic precision for site planning and progress tracking.

- **Infrastructure Management**: Pix4Dinspect streamlines fault detection and maintenance for assets like power lines.
- **Agriculture**: Pix4Dfields provides actionable crop health insights to optimize farming decisions.
- **Advanced Mapping**: Pix4Dmatic fuses photogrammetry with LiDAR data for high-accuracy environmental and urban mapping.

Rich Array of Deliverables

The suite produces a versatile set of outputs:

- **Orthomosaic Maps**: High-resolution, georeferenced maps for planning and agricultural analysis.
- **3D Models**: Detailed visualizations supporting real estate and cultural heritage projects.
- **Videos and Photos**: Professional-grade media for documentation and presentations.
- **Panoramic Views**: 360-degree imagery enhancing site inspections and virtual tours.

Advantages of a Web-Based Platform

Pix4D's cloud integration elevates its utility:

- **Global Access**: Pix4Dcloud enables remote project management and real-time team collaboration.
- **Scalability**: Processes large datasets efficiently, reducing reliance on local hardware.
- **Continuous Improvement**: Automatic updates ensure access to cutting-edge features.

Thoughts

Pix4D stands out as a powerhouse in drone photogrammetry, blending specialized tools, robust GIS integration, and cloud flexibility. Its tailored variants and comprehensive outputs make it indispensable across industries. As aerial mapping technology progresses, Pix4D is poised to remain a leader in driving precision and innovation.

Maps Made Easy

Maps Made Easy (MME) is a straightforward, cloud-based platform designed to process drone imagery into custom maps and models, targeting general users in agriculture, real estate, mining, and construction monitoring. This review assesses its core features, practical applications, and notable limitations, particularly its constraints for professional-grade geospatial tasks.

Key Features and Capabilities

MME prioritizes simplicity with automated tools:

- **Image Processing**: Automatically stitches uploaded images into orthomosaic maps, streamlining basic output creation.
- **Point Cloud and 3D Modeling**: Generates rudimentary point clouds and 3D models, though with less detail than advanced competitors.

- **Volume Measurement**: Calculates stockpile volumes, supporting construction and mining workflows.
- **Vegetation Analysis**: Produces NDVI maps to assess plant health for agricultural use.

Industry-Specific Applications

MME serves modest, small-scale needs:

- **Agriculture**: NDVI outputs aid in monitoring crop conditions for hobbyists or small farms.
- **Real Estate**: Provides simple maps and models for property visualization.
- **Construction and Mining**: Offers basic volume estimates for informal site tracking.

Rich Array of Deliverables

The platform delivers functional, if limited, outputs:

- **Orthomosaic Maps**: Georeferenced maps suitable for quick visual assessments.
- **3D Models**: Basic visualizations for casual planning or presentation.
- **Point Clouds**: Simplified datasets for elementary elevation analysis.

Advantages and Limitations

While MME leverages cloud-based accessibility, it falls short for advanced use:

- **Ease of Use**: Its intuitive design caters to beginners needing fast results without complexity.
- **GIS Compatibility**: Exports GeoTIFF files, but lacks robust integration with tools like ArcGIS or QGIS.
- **Key Constraints**:
    - *Accuracy*: No Ground Control Point (GCP) support compromises precision, unfit for legal or survey-grade work.
    - *Customization*: Limited processing options restrict flexibility compared to Pix4D or DroneDeploy.
    - *Scalability*: Struggles with large datasets or intricate projects, hindering professional applications.

Thoughts

Maps Made Easy excels as an accessible, entry-level solution for quick, small-scale mapping tasks. Its automated features and cloud-based simplicity appeal to casual users, but its lack of GCP support, limited customization, and reduced scalability render it inadequate for survey-grade or complex professional needs. For those requiring precision and depth, specialized platforms remain the superior choice.

Agisoft Metashape Pro

Agisoft Metashape Professional is a premier photogrammetry solution that transforms digital imagery into high-precision 3D models and geospatial data, serving professionals in surveying, cultural heritage,

GIS, and visual effects. Renowned for its robust processing engine and versatile capabilities, this review examines its advanced features, industry applications, and integration strengths, highlighting its role as a cornerstone in drone-based mapping workflows.

Key Features and Capabilities

Metashape Professional offers a sophisticated toolkit for photogrammetric excellence:

- **Photogrammetry and LiDAR Fusion**: Processes aerial and close-range imagery alongside LiDAR data, delivering detailed point clouds and models with exceptional accuracy.
- **Georeferencing Precision**: Supports Ground Control Points (GCPs), RTK/PPK data, and geoid models (e.g., EGM96, EGM2008) for survey-grade spatial alignment.
- **Automation and Customization**: Combines automated workflows with scripting (Python, Global Mapper Script) for tailored processing and scalability.
- **Multispectral Support**: Handles multispectral and thermal imagery, enabling advanced vegetation and environmental analysis.

Industry-Specific Applications

The software adapts seamlessly to diverse professional demands:

- **Surveying**: Produces accurate topographic maps and elevation models for land management and infrastructure planning.
- **Cultural Heritage**: Generates detailed 3D reconstructions for documentation and preservation of archaeological sites.
- **Construction and Mining**: Calculates volumes and monitors site progress with precision for operational efficiency.
- **Environmental Monitoring**: Analyzes multispectral data for ecosystem studies and change detection.

Rich Array of Deliverables

Metashape Professional delivers a comprehensive set of outputs:

- **Orthomosaic Maps**: High-resolution, georeferenced maps for planning and GIS integration.
- **3D Models**: Textured, photorealistic models for visualization and analysis.
- **Point Clouds**: Dense, editable datasets for elevation and volumetric assessments.
- **Digital Elevation Models (DEMs)**: Detailed terrain representations for engineering and environmental use.

Advantages of Versatility and Precision

Its strengths are tempered by practical considerations:

- **Processing Power**: Leverages GPU acceleration and multi-core CPUs for rapid handling of large datasets, outpacing many competitors in speed.

- **GIS Integration**: Exports LAS/LAZ point clouds, GeoTIFF orthomosaics, and other formats, compatible with ArcGIS, QGIS, and CAD platforms.
- **Cost-Effectiveness**: Offers a perpetual license at $3,499 (with educational discounts dropping to $59-$549), providing value over subscription-based alternatives.
- **Challenges**: A steep learning curve and high hardware demands (e.g., 32GB+ RAM, NVIDIA/AMD GPU) may challenge new users or those with limited resources.

Thoughts

Agisoft Metashape Professional stands out as a powerhouse in photogrammetry, blending speed, precision, and versatility to meet the needs of surveyors, heritage specialists, and geospatial professionals. Its ability to fuse LiDAR and photogrammetric data, coupled with robust GIS integration, positions it as a top-tier choice for complex projects. While its technical demands may require an investment in hardware and expertise, its perpetual licensing and expansive capabilities ensure long-term value, making it a vital tool as drone mapping continues to evolve.

Propeller Aero

Propeller Aero delivers an end-to-end drone data solution tailored for construction, mining, aggregates, and waste management, blending ease of use with high-precision outputs. This review explores its photogrammetric capabilities, cloud-based infrastructure, and advanced analytics, highlighting its value for professional mapping and surveying workflows.

Key Features and Capabilities

Propeller Aero combines robust tools for aerial data processing:

- **Photogrammetry**: Transforms drone imagery into accurate 2D maps and 3D models, supporting detailed site analysis.
- **AeroPoints Integration**: Employs smart Ground Control Points (GCPs) to achieve survey-grade positional accuracy.
- **Advanced Analytics**: Provides volume calculations and progress tracking for operational insights.
- **High Compatibility**: Supports a wide range of drones and cameras, enhancing flexibility.

Industry-Specific Applications

The platform excels in sector-specific tasks:

- **Construction**: Generates precise models and maps for site planning and progress monitoring.
- **Mining and Aggregates**: Measures stockpile volumes and ensures compliance with regulatory standards.
- **Waste Management**: Tracks landfill changes and supports environmental assessments.
- **Surveying**: Delivers reliable topographic data for land development and resource management.

Rich Array of Deliverables

Propeller Aero produces actionable outputs:

- **Orthomosaic Maps**: High-resolution, georeferenced maps for planning and documentation.
- **3D Models**: Detailed visualizations for project estimation and virtual oversight.
- **Analytical Reports**: Volume and progress data for operational decision-making.

Advantages of a Web-Based Platform

Its cloud infrastructure drives efficiency:

- **Global Access**: Enables remote project management and real-time team collaboration.
- **Scalability**: Adapts seamlessly to projects of varying size, from small sites to expansive operations.
- **Time Savings**: Streamlines processing and data sharing, accelerating workflows.
- **Continuous Improvement**: Regular updates keep tools aligned with industry needs.

Thoughts

Propeller Aero stands out as a top-tier solution for professionals in mapping and surveying, merging survey-grade accuracy with cloud-based efficiency. Its integration of AeroPoints, advanced analytics, and broad compatibility makes it a reliable choice for construction, mining, and beyond. As drone-driven insights grow in demand, Propeller Aero is well-equipped to enhance precision and productivity across industries.

Global Mapper Pro

Global Mapper Pro is a sophisticated GIS tool excelling in spatial data processing, analysis, and visualization, tailored for surveying, engineering, and urban planning. This review highlights its versatile feature set, intuitive design, and cost-effective performance, underscoring its value for geospatial professionals.

Key Features and Capabilities

Global Mapper Pro offers a robust toolkit for spatial workflows:

- **Data Integration**: Supports over 300 file formats, enabling seamless import and export of diverse datasets.
- **3D Visualization**: Generates detailed terrain models and interactive fly-throughs for enhanced analysis.
- **Spatial Processing**: Includes geoprocessing and LiDAR tools for advanced data manipulation.
- **Cartographic Tools**: Produces customizable, high-quality maps for professional outputs.
- **Automation**: Features scripting in Global Mapper Script and Python to streamline repetitive tasks.

Industry-Specific Applications

The software adapts to a range of professional uses:

- **Surveying**: Delivers precise mapping and boundary delineation for land development.

- **Environmental Analysis**: Supports habitat mapping and impact assessments with detailed spatial insights.
- **Urban Planning**: Visualizes sites and infrastructure for design and decision-making.

Rich Array of Deliverables

Global Mapper Pro provides versatile outputs:

- **Terrain Models**: 3D representations for elevation and topographic studies.
- **Custom Maps**: Georeferenced cartographic products for planning and reporting.
- **Visualization Files**: Fly-throughs and spatial datasets for presentations and analysis.

Advantages of Versatility and Design

Its strengths lie in accessibility and efficiency:

- **Broad Compatibility**: Handles diverse data types, from raster to vector, with ease.
- **User-Friendly Interface**: Balances advanced functionality with an intuitive layout, appealing to novices and experts alike.
- **Performance**: Processes large datasets efficiently, even on modest hardware.
- **Affordability**: Offers premium GIS capabilities at a competitive price point.

Thoughts

Global Mapper Pro stands out as a comprehensive and cost-effective GIS solution, blending powerful spatial tools with an approachable design. Its versatility and performance make it a valuable asset for surveyors, engineers, and planners. As geospatial demands grow, Global Mapper Pro remains a reliable choice for delivering actionable insights.

ArcGIS Drone2Map

ArcGIS Drone2Map, developed by Esri, transforms drone-captured data into high-precision maps and models, serving agriculture, construction, and public safety. Integrated with the ArcGIS ecosystem, it offers advanced geospatial capabilities. This review explores its processing power, analytics, and seamless compatibility, emphasizing its role in professional mapping and surveying.

Key Features and Capabilities

ArcGIS Drone2Map provides a streamlined drone-to-GIS pipeline:

- **Drone Data Processing**: Automates workflows to convert imagery into detailed maps and models.
- **ArcGIS Integration**: Enhances analysis and updates within the broader ArcGIS platform.
- **Advanced Analytics**: Supports 3D modeling, orthomosaic generation, and detailed reporting.
- **Broad Compatibility**: Works with various drone systems and exports to multiple formats.

Industry-Specific Applications

The software excels across key sectors:

- **Agriculture**: Produces maps for crop monitoring and resource optimization.
- **Construction**: Supplies topographic data and 3D models for site management.
- **Public Safety**: Enables rapid mapping for emergency planning and response.
- **Surveying**: Delivers precise geospatial outputs for land and infrastructure projects.

Rich Array of Deliverables

ArcGIS Drone2Map generates professional-grade outputs:

- **Orthomosaic Maps**: High-resolution, georeferenced maps for planning and analysis.
- **3D Models**: Accurate visualizations for design and operational oversight.
- **Analytical Reports**: Detailed metrics for project tracking and decision support.

Advantages of Integration and Precision

Its Esri-backed design offers distinct benefits:

- **High Accuracy**: Ensures reliable, survey-grade results for critical applications.
- **Efficiency**: Streamlines data processing and GIS workflows, saving time.
- **Scalability**: Adapts to projects of any size, from small plots to large regions.
- **User-Friendly**: Combines intuitive controls with customizable templates for accessibility.

Thoughts

ArcGIS Drone2Map stands out as a leader in drone mapping, leveraging its deep integration with ArcGIS to deliver precision and versatility. Its advanced analytics and broad compatibility make it an essential tool for professionals in surveying, construction, and beyond. As drone technology advances, Drone2Map is well-positioned to drive geospatial innovation and efficiency.

Virtual Surveyor

Virtual Surveyor bridges drone-captured data and CAD-ready topographic outputs, serving construction, mining, and agriculture with a focus on actionable survey results. Lets explore its terrain processing, survey tools, and CAD/GIS integration, highlighting its efficiency for professional workflows.

Key Features and Capabilities

Virtual Surveyor offers targeted tools for survey-grade outputs:

- **Data Processing**: Imports orthophotos, Digital Surface Models (DSMs), and point clouds for precise terrain editing.
- **Survey Tools**: Calculates volumes and generates contours for detailed land analysis.
- **CAD/GIS Integration**: Exports DXF/DWG files and supports GIS platforms for seamless compatibility.

Industry-Specific Applications

The software excels in practical use cases:

- **Construction**: Produces topographic maps for site planning and earthwork calculations.
- **Mining**: Supports volume assessments and terrain monitoring for operational efficiency.
- **Agriculture**: Delivers land data for drainage and field management.

Rich Array of Deliverables

Virtual Surveyor generates versatile survey outputs:

- **Topographic Maps**: Detailed, georeferenced maps for planning and analysis.
- **3D Terrain Models**: Editable representations for design and visualization.
- **Contour Lines**: Accurate elevation data for engineering and land use.

Advantages of Usability and Integration

Its design enhances productivity:

- **User-Friendly Interface**: Intuitive controls and customizable workflows simplify complex tasks.
- **Collaboration**: Cloud-based features enable real-time teamwork and data sharing.
- **Precision**: Ensures accurate terrain edits for reliable survey results.

Thoughts

Virtual Surveyor stands out as an efficient bridge between drone data and CAD-ready deliverables, offering precision and usability for survey professionals. Its strong integration with CAD and GIS workflows makes it a valuable tool for construction, mining, and beyond. As drone surveying grows, Virtual Surveyor is well-positioned to streamline topographic analysis.

RealityCapture 1.4

RealityCapture 1.4, developed by Capturing Reality, is a premier photogrammetry software renowned for its speed and precision in creating 3D models and maps. Originally designed for video games, it now shines in surveying and CGI. This review assesses its processing power, geodetic features, and trade-offs, highlighting its standout performance.

Key Features and Capabilities

RealityCapture combines speed with advanced functionality:

- **Photogrammetry and LiDAR**: Merges imagery and laser scans for highly accurate 3D outputs.
- **Geoid Support**: Incorporates EGM96, EGM2008, and custom geoids for precise altitude mapping.
- **Editing Tools**: Refines point clouds and meshes for detailed customization.
- **Broad Compatibility**: Supports a wide range of cameras and drones for flexible input.

Industry-Specific Applications

The software adapts to professional and creative needs:

- **Surveying**: Delivers precise topographic models for land and infrastructure projects.
- **CGI and Visualization**: Creates photorealistic 3D assets for film and gaming.
- **Construction**: Supports detailed site analysis and progress tracking.

Rich Array of Deliverables

RealityCapture produces high-fidelity outputs:

- **3D Models**: Textured, accurate models for visualization and analysis.
- **Orthomosaic Maps**: Georeferenced maps for spatial planning.
- **Point Clouds**: Dense datasets for elevation and structural studies.

Advantages and Limitations

Its strengths come with caveats:

- **Unmatched Speed**: Processes thousands of images faster than most competitors.
- **Precision**: Combines photogrammetry and LiDAR for exceptional accuracy.
- **Scalability**: Adapts to large datasets and complex projects.
- **Challenges**: A steep learning curve and high hardware requirements may deter beginners.

Thoughts

RealityCapture 1.4 stands out as a top-tier photogrammetry tool, offering unparalleled speed and precision for 3D modeling and mapping. While its complexity and resource demands pose challenges, its performance makes it a preferred choice for surveyors, CGI artists, and professionals tackling intricate projects. As photogrammetry evolves, RealityCapture remains a benchmark for quality and efficiency.

Summary

This chapter has evaluated an extensive lineup of photogrammetry software platforms, each designed to convert drone-captured imagery into precise geospatial deliverables. From DJI Terra to RealityCapture, these tools address the diverse needs of industries such as construction, agriculture, surveying, and environmental monitoring. Platforms like Pix4D and DroneDeploy excel in professional-grade versatility, while Maps Made Easy offers streamlined simplicity for casual users, illustrating the broad spectrum of capabilities available.

The progression of these solutions underscores the deepening integration of drone technology with geospatial analysis. As drones evolve, equipped with enhanced sensors and superior flight performance, photogrammetry software continues to advance, incorporating innovations like artificial intelligence for automated feature recognition and robust cloud computing for real-time collaboration. This trajectory promises heightened accuracy, efficiency, and applicability, extending the reach of aerial data into emerging domains such as urban planning and disaster response. By leveraging the distinct strengths and navigating the limitations of each platform, professionals can optimize these tools to meet specific project requirements, unlocking the full potential of drone driven insights within an increasingly sophisticated digital landscape.

Chapter 8. LiDAR Software

LiDAR (Light Detection and Ranging) and photogrammetry are both powerful technologies used in drone mapping and spatial data collection. Still, they operate on fundamentally different principles and offer distinct advantages and challenges.

LiDAR Software Overview

LiDAR is a remote sensing technology that uses laser pulses to measure distances to a target. These pulses are emitted from a LiDAR sensor, bounce off objects or surfaces, and return to the sensor, allowing the software to calculate the distance based on the time it takes for the pulses to return. By rapidly emitting thousands or millions of pulses per second, LiDAR can generate highly detailed and accurate 3D point clouds that represent the physical world in precise detail. In some scenarios

LiDAR software processes the raw data captured by the LiDAR sensors, transforming it into actionable 3D models, terrain maps, and other geospatial datasets. This software is crucial for interpreting the complex datasets that LiDAR produces, often integrating with Geographic Information Systems (GIS) and other spatial analysis tools. LiDAR is particularly valued in industries that require high precision and the ability to penetrate vegetation or other obstructions to reveal the underlying terrain.

Comparing and Contrasting LiDAR with Photogrammetry

1. **Data Acquisition:**

   o **Photogrammetry** relies on capturing multiple overlapping photographs of an area using high-resolution cameras. The software then uses these images to generate 2D and 3D models through a process called Structure from Motion (SfM).

   o **LiDAR**, on the other hand, uses laser pulses to measure the distance to objects directly, creating a point cloud that represents the environment's 3D structure. Unlike photogrammetry, LiDAR can capture data in low-light conditions or through obstacles like dense vegetation.

2. **Accuracy and Precision:**

   o **Photogrammetry** can achieve high accuracy, particularly in well-lit conditions and with sufficient image overlap. However, its precision can be affected by factors like camera quality, altitude, and environmental conditions. Photogrammetry is typically more suitable for applications where relative accuracy (i.e., the relationship between points) is more important than absolute accuracy.

   o **LiDAR** offers superior precision and is often preferred for tasks requiring absolute accuracy, such as topographic mapping, forestry, and infrastructure inspection. It is particularly effective in environments where photogrammetry struggles, such as heavily vegetated areas, where it can penetrate through foliage to map the ground beneath.

3. **Data Processing:**

   o **Photogrammetry software** uses sophisticated algorithms to stitch images together, generate point clouds, and create 3D models or orthomosaics. While computationally intensive, this process is highly automated and produces visually detailed models ideal for construction monitoring, urban planning, and agriculture applications.

- o **LiDAR software** focuses on processing the raw point cloud data generated by the LiDAR sensors. This software often includes tools for filtering out noise, classifying points (e.g., ground vs. vegetation), and integrating the data with other geospatial datasets. LiDAR data processing can be more complex and requires specialized software to handle large datasets and achieve the desired level of detail.

4. **Applications:**

- o **Photogrammetry** is widely used in industries like construction, mining, agriculture, and archaeology due to its ability to create detailed and accurate 3D models from easily acquired photographic data. It is especially effective in environments where visual detail and color information are important.

- o **LiDAR** is preferred in applications where precision and the ability to see through obstructions are critical. This includes forestry, floodplain mapping, infrastructure inspection, and archaeological site discovery. LiDAR's ability to accurately map the terrain beneath tree canopies makes it invaluable in environmental monitoring and natural resource management.

5. **Cost and Accessibility:**

- o **Photogrammetry** is generally more accessible and cost-effective than LiDAR. The equipment required (drones with high-resolution cameras) is widely available, and the software is often less expensive and more user-friendly.

- o While offering higher precision, LiDAR involves higher costs due to the required specialized sensors and processing software. The investment is justified in projects where accuracy and the ability to capture data in challenging conditions are paramount.

Both LiDAR and photogrammetry have their strengths and are best suited to different types of projects. Photogrammetry is ideal for scenarios where visual detail and lower costs are important, while LiDAR excels in environments requiring high precision and the ability to map through obstructions. Understanding the differences between these technologies helps professionals choose the right tool for their specific needs, whether creating detailed topographic maps, managing forest resources, or conducting infrastructure inspections.

In the following sections, we will explore some of the leading LiDAR software platforms in the industry, each offering unique features tailored to various applications. This will provide a deeper understanding of how LiDAR complements photogrammetry and the specific scenarios where it is preferred.

DJI Terra
Overview: DJI Terra 3.4.5

DJI Terra 3.4.5 is an advanced LiDAR data processing and mapping software designed to work seamlessly within DJI's ecosystem. Primarily aimed at construction, surveying, and environmental monitoring industries, this software enables users to process aerial LiDAR data captured by DJI drones into detailed 3D models, point clouds, and other geospatial outputs. With the growing importance of LiDAR technology in capturing high-precision data, DJI Terra 3.4.5 has evolved to be a powerful tool for professionals who need accurate, real-time mapping and modeling capabilities.

Description of DJI Terra 3.4.5

DJI Terra 3.4.5 is specifically optimized to process LiDAR data, offering advanced features that enhance the accuracy and detail of the models generated. The software is designed to handle the high-density point clouds produced by LiDAR sensors, enabling users to create highly detailed 3D reconstructions of their environments. DJI Terra supports integrating LiDAR data with other sensor data, such as RGB images, producing more comprehensive models.

One of the standout features of DJI Terra 3.4.5 is its ability to process and visualize large LiDAR datasets efficiently. The software can handle the immense data volumes typical of LiDAR scans, ensuring users can generate accurate models without compromising speed. Additionally, DJI Terra supports various coordinate systems and geoid models, allowing for precise georeferencing, which is critical in professional surveying and mapping projects.

Functionality and Features

- **LiDAR-Specific Data Processing**: DJI Terra 3.4.5 excels at processing LiDAR data captured by DJI drones. It allows users to generate dense, accurate point clouds crucial for detailed terrain analysis and 3D modeling. The software can handle the raw LiDAR data and convert it into actionable insights with minimal manual intervention.

- **Real-Time Point Cloud Visualization**: The software allows users to visualize LiDAR point clouds in real-time during data capture. This feature is particularly beneficial for field operations, as it enables immediate quality checks and adjustments, ensuring that the collected data meets the project's requirements.

- **Multi-Sensor Integration**: DJI Terra supports the fusion of LiDAR data with photogrammetric data from RGB cameras, enhancing the detail and realism of the generated models. This multi-sensor integration is essential for creating comprehensive 3D maps with structural details and accurate terrain representations.

- **High-Resolution Digital Surface Models (DSM) and Digital Terrain Models (DTM)**: DJI Terra can generate DSMs and DTMs from LiDAR data, providing critical information for applications such as flood modeling, land-use planning, and environmental management. These models offer precise topographical details, which are indispensable for various geospatial analyses.

- **Support for Complex Terrain**: The software is particularly effective in processing LiDAR data from complex terrains, such as forests, urban areas, and steep landscapes. The ability to penetrate vegetation and capture ground-level details makes it ideal for forestry management, archaeology, and infrastructure monitoring.

- **Advanced Georeferencing**: DJI Terra 3.4.5 includes robust georeferencing tools that allow users to apply various coordinate systems and geoid models, ensuring that LiDAR data is accurately aligned with real-world locations. This capability is crucial for ensuring the precision of models used in surveying and civil engineering projects.

Evaluation of Functionality and Usefulness

Advantages:

- **Optimized for LiDAR Data**: DJI Terra 3.4.5 is designed to process LiDAR data, making it an excellent choice for users requiring high-density point clouds and detailed terrain models. Its ability to handle large datasets efficiently ensures users can work with complex projects without compromising speed or accuracy.

- **Seamless DJI Integration**: The software's seamless integration with DJI drones and sensors simplifies the workflow from data capture to processing, making it easier for users to manage their projects without needing multiple software platforms.

- **Real-Time Processing and Visualization**: DJI Terra's real-time capabilities allow for immediate data validation in the field, reducing the likelihood of errors and the need for repeat missions.

Limitations:

- **Hardware Dependency**: While DJI Terra is powerful, it is heavily dependent on DJI hardware, which may limit its flexibility for users who work with a broader range of equipment or need to integrate data from non-DJI sources.

- **Learning Curve for Advanced Features**: The software's advanced LiDAR processing features can present a learning curve, particularly for users new to LiDAR technology or who have primarily worked with photogrammetry.

- DJI Terra 3.4.5 is a robust and highly capable software solution for professionals working with LiDAR data within the DJI ecosystem. Its ability to process and visualize LiDAR data in real time and its seamless integration with DJI drones make it an invaluable tool for industries that require precise 3D mapping and terrain modeling. Despite its limitations regarding hardware flexibility and the learning curve for advanced features, DJI Terra 3.4.5 remains a top choice for users seeking a comprehensive LiDAR data processing tool that delivers accuracy, efficiency, and ease of use.

Rock Robotic
Overview

Rock Robotic provides a comprehensive LiDAR solution that integrates hardware and software and is designed specifically for surveying, construction, and infrastructure inspection professionals. The Rock R360Pro LiDAR hardware and Rock Robotic's processing software offer a seamless workflow from data capture to final analysis. This end-to-end solution is tailored to deliver high-precision results, making it ideal for applications where accuracy and detail are paramount.

Description of Rock Robotic Solution

Rock Robotic's LiDAR solution is centered around the Rock R3Pro LiDAR system, a high-precision sensor that captures detailed 3D data across various environments. The hardware is lightweight and versatile, making it suitable for various deployment methods, including drones, vehicles, and handheld devices. This flexibility ensures that the Rock R3Pro can be utilized effectively in different terrains and scenarios, from dense urban areas to remote landscapes.

The accompanying software suite is designed to process the raw LiDAR data captured by the R1A system. It provides powerful tools for point cloud processing, ground classification, and 3D modeling. The software's capabilities extend to integration with GIS platforms, allowing users to incorporate LiDAR data into larger geospatial projects seamlessly. This integration is particularly beneficial for professionals working in land surveying, urban planning, and infrastructure development, where LiDAR data can be used to create accurate maps and models.

Functionality and Features

1. **Point Cloud Processing**: The Rock Robotic software excels in processing the dense point clouds generated by the Rock R1A LiDAR system. It offers tools for filtering, classification, and segmentation, allowing users to extract meaningful information from the raw data. The software can automatically distinguish between ground surfaces, vegetation, and man-made structures, which is essential for creating detailed terrain models and identifying features of interest.

2. **Ground Classification**: One critical feature of the Rock Robotic software is its ability to classify ground points accurately. This is particularly useful in applications like topographic mapping and construction site monitoring, where understanding the ground surface is crucial. The software uses advanced algorithms to separate ground points from other elements in the LiDAR data, ensuring that the resulting models are both accurate and reliable.

3. **3D Modeling**: The software allows users to create detailed 3D models from the processed LiDAR data. These models can be used for various purposes, including infrastructure inspection, volumetric analysis, and virtual simulations. The high level of detail provided by the Rock R360Pro LiDAR system ensures that these models are highly accurate, making them valuable tools for decision-making in construction and surveying projects.

4. **Integration with GIS Platforms**: Rock Robotic's software is designed to integrate seamlessly with popular GIS platforms, enabling users to incorporate LiDAR data into their broader geospatial workflows. This integration is particularly beneficial for professionals who need to analyze LiDAR data in conjunction with other types of geospatial information, such as satellite imagery or cadastral maps.

Evaluation of Functionality and Usefulness

Advantages:

- **High Accuracy**: The Rock R1A LiDAR system, combined with Rock Robotic's processing software, delivers highly accurate results, making it ideal for applications where precision is critical.

- **Comprehensive Solution**: By offering both hardware and software, Rock Robotic provides a complete solution that covers the entire workflow from data capture to analysis. This ensures compatibility and optimization at every stage of the process.

- **User-Friendly Interface**: Despite its advanced capabilities, the Rock Robotic software is designed with a user-friendly interface, making it accessible to professionals without extensive LiDAR technology experience.

Limitations:

- **Hardware Dependency**: While the software is highly effective, it is primarily designed to work with the Rock R1A LiDAR system. This dependency on specific hardware may limit flexibility for users who need to integrate data from other LiDAR systems or prefer alternative equipment.

- **Investment Cost**: Investing in both the Rock R1A hardware and the accompanying software can be a significant upfront cost, particularly for smaller operations or those with existing investments in other LiDAR solutions.

Rock Robotic provides a powerful, all-in-one LiDAR solution that combines high-precision hardware with advanced processing software. This makes it an excellent choice for surveying, construction, and infrastructure inspection professionals who require reliable and detailed 3D data. While the solution's reliance on specific hardware may be a drawback for some users, its overall accuracy, comprehensive functionality, and ease of use make it a compelling option for those seeking an integrated LiDAR system.

Phoenix LiDAR Systems
Overview

Phoenix LiDAR Systems is a leading provider of advanced LiDAR solutions, offering hardware and software for high-resolution mapping and surveying. The company's software suite is specifically tailored to handle the complex data generated by their LiDAR systems, providing tools for data visualization, automated processing workflows, and in-depth analysis of point clouds. Phoenix LiDAR Systems' solutions are highly versatile, making them suitable for various applications, including topographic mapping, forestry management, and infrastructure monitoring.

Description of Phoenix LiDAR Systems

Phoenix LiDAR Systems offers a range of LiDAR sensors known for their high resolution and accuracy. These sensors are designed to be deployed on various platforms, including drones, vehicles, and backpacks, allowing for flexible data collection in diverse environments. The hardware is complemented by a sophisticated software suite that enables users to easily process, analyze, and visualize the LiDAR data.

The software is designed to handle large datasets efficiently, allowing users to process complex point clouds and generate detailed 3D models. Phoenix LiDAR Systems strongly emphasizes automation, with tools that streamline the data processing workflow, reducing the time and effort required to generate accurate results. The software also supports a wide range of LiDAR sensors, including those from other manufacturers, which adds to its versatility.

Functionality and Features

1. **Data Visualization**: Phoenix LiDAR Systems' software includes advanced visualization tools that allow users to explore and interact with LiDAR point clouds in 3D. This feature is crucial for

understanding the data's spatial relationships and identifying key features such as vegetation, buildings, and other structures. Visualizing data in real time also aids in quality control during data collection.

2. **Automated Processing Workflows**: The software offers automated workflows that significantly reduce the manual effort required to process LiDAR data. This includes automatic point cloud classification, ground extraction, and feature detection. These automation tools are designed to handle large volumes of data quickly and efficiently, making the software particularly useful for large-scale projects where time is of the essence.

3. **Support for Various LiDAR Sensors**: Phoenix LiDAR Systems' software is compatible with a wide range of LiDAR sensors, not just those produced by Phoenix. This flexibility allows users to integrate data from multiple sources, making the software suitable for projects that require different types of LiDAR data or that involve collaborations with other organizations.

4. **Advanced Analysis Tools**: The software includes a suite of analysis tools that allow users to perform detailed examinations of their LiDAR data. This includes volumetric analysis, contour generation, and vegetation mapping. These tools are essential for forestry, agriculture, and environmental monitoring applications, where precise measurements and detailed data are required.

Evaluation of Functionality and Usefulness

Advantages:

- **High-Resolution Data Processing**: Phoenix LiDAR Systems' software is designed to handle the high-resolution data generated by their LiDAR sensors, ensuring that users can extract the maximum detail from their point clouds.

- **Versatile Use Cases**: The software's support for various LiDAR sensors and its broad range of applications make it a versatile tool for professionals in many industries.

- **Advanced Automation Features**: The software's automated workflows save time and reduce the potential for errors in data processing, making it ideal for large-scale projects.

Limitations:

- **Hardware Focus**: While the software is highly capable, it is primarily designed to work with Phoenix LiDAR hardware. This focus may limit its appeal to users who prefer LiDAR systems from other manufacturers or who have existing investments in other LiDAR technologies.

- **Cost Considerations**: Phoenix LiDAR Systems' solutions, including both hardware and software, are generally positioned at the higher end of the market. This can be a significant investment, particularly for smaller companies or those just entering the field of LiDAR-based surveying.

Phoenix LiDAR Systems provides a robust and versatile solution for professionals needing advanced LiDAR data processing and analysis. The combination of high-resolution hardware and powerful, automated software workflows makes it a top choice for industries ranging from topographic mapping to environmental monitoring. While the software's close integration with Phoenix's hardware may be a limitation for some, its overall flexibility, high-quality output, and time-saving features make it an invaluable tool for those looking to maximize the potential of their LiDAR data.

LiDAR360
Overview

LiDAR360, developed by GreenValley International, is a comprehensive software suite designed for processing and analyzing LiDAR data across a range of industries. It is particularly well-regarded in fields such as forestry, powerline inspection, and urban modeling, where its specialized tools and robust processing capabilities make it an indispensable resource. LiDAR360 offers an extensive set of features tailored to meet the specific needs of these sectors, making it a versatile and powerful tool for professionals engaged in large-scale LiDAR data analysis.

Description of LiDAR360

LiDAR360 is engineered to handle large and complex LiDAR datasets, focusing on delivering precise and actionable insights. The software is designed to facilitate the entire workflow, from raw data processing to detailed analysis and model creation. Its suite of tools includes advanced algorithms for point cloud classification, terrain analysis, and vegetation modeling, which are particularly beneficial for applications in forestry and environmental monitoring.

One of the software's standout features is its ability to perform canopy height modeling, making it a preferred choice for forestry professionals who need to assess forest structure and biomass. Additionally, LiDAR360 offers powerful tools for urban modeling, enabling the accurate reconstruction of urban environments and infrastructure, which is critical for urban planning and development projects. The software also excels in powerline inspection, where it provides specialized tools for analyzing the clearance of vegetation and other potential hazards near powerlines.

Functionality and Features

1. **Point Cloud Classification**: LiDAR360 provides advanced point cloud classification tools that automatically categorize LiDAR data into different classes, such as ground, vegetation, and buildings. This capability is essential for users who need to extract specific features from their LiDAR data and is particularly useful in forestry and urban modeling, where detailed classification is necessary for accurate analysis.

2. **Terrain Analysis**: The software includes a comprehensive set of tools for terrain analysis, allowing users to create detailed Digital Elevation Models (DEMs) and perform slope and aspect analysis. These features are critical for applications in land management, infrastructure development, and environmental monitoring, where understanding the terrain is vital for decision-making.

3. **Canopy Height Modeling**: One of LiDAR360's unique features is its ability to model canopy height with high precision. This tool is especially useful in forestry, where understanding the vertical structure of forests is essential for biomass estimation, habitat assessment, and forest management. The software can generate detailed canopy height models (CHMs) that provide insights into forest health and growth patterns.

4. **Urban Modeling**: LiDAR360 offers powerful urban modeling tools that enable users to create accurate 3D models of urban environments. These tools are crucial for urban planning,

infrastructure assessment, and disaster management. The software allows for the precise reconstruction of buildings, roads, and other urban features, facilitating the analysis and simulation of urban growth and development.

5. **Powerline Inspection**: The software includes specialized tools for powerline inspection, enabling users to analyze the clearance between vegetation and powerlines. This is critical for preventing outages and ensuring the safety of power infrastructure. LiDAR360's ability to automatically detect potential hazards makes it a valuable tool for utility companies and contractors involved in vegetation management and infrastructure maintenance.

Evaluation of Functionality and Usefulness

Advantages:

- **Specialized Tools for Forestry and Powerline Analysis**: LiDAR360's targeted tools for canopy height modeling and powerline inspection make it highly effective for professionals in forestry and utilities. These features are tailored to meet the specific needs of these industries, providing accurate and actionable insights.

- **Robust Processing Capabilities**: The software is designed to handle large and complex datasets efficiently, making it suitable for large-scale projects that require detailed analysis and modeling. Its robust processing capabilities ensure that users can work with high-density point clouds without performance issues.

- **Cost-Effective**: Compared to other high-end LiDAR processing software, LiDAR360 is relatively cost-effective, making it accessible to a broader range of users, including smaller firms and research institutions.

Limitations:

- **Steeper Learning Curve**: LiDAR360's advanced features and specialized tools may present a steeper learning curve for users who are not familiar with LiDAR data processing. This can require additional training or time investment to utilize the software's capabilities fully.

- **Limited to Specific Applications**: While LiDAR360 excels in forestry, powerline inspection, and urban modeling, it may not be as versatile for other applications outside these domains. Users in fields that require more generalized LiDAR processing tools may find the software less applicable to their needs.

LiDAR360 by GreenValley International is a powerful and versatile software suite designed to meet the needs of professionals in forestry, powerline inspection, and urban modeling. Its specialized tools for point cloud classification, terrain analysis, and canopy height modeling make it an essential resource for those working in these fields. While the software's advanced capabilities can present a learning curve, its robust processing power and cost-effectiveness make it a valuable tool for large-scale LiDAR data analysis. For professionals seeking detailed and accurate insights into vegetation, terrain, and urban environments, LiDAR360 provides the tools necessary to deliver high-quality results.

Terrasolid
Overview

Terrasolid is an industry-leading software suite designed to process airborne, mobile, and terrestrial LiDAR data. Known for its robust capabilities and precision, Terrasolid has become a go-to solution for urban planning, infrastructure management, and environmental analysis professionals. Its ability to handle large datasets efficiently while delivering detailed and accurate outputs makes it a preferred choice for complex projects requiring high-resolution LiDAR data. With its extensive feature set and strong community support, Terrasolid is recognized as a standard in the field of LiDAR data processing.

Description of Terrasolid

Terrasolid's software suite is renowned for its versatility and precision in processing various types of LiDAR data. Whether the data is captured from airborne platforms, mobile units, or terrestrial systems, Terrasolid provides a comprehensive set of tools for processing and analyzing these datasets. The software excels in converting raw LiDAR point clouds into actionable information that can be used in a wide range of applications, from building extraction and ground surface modeling to vegetation analysis and urban infrastructure management.

The software is highly regarded for its accuracy, especially in point cloud classification, where it can distinguish between different types of surfaces, such as ground, buildings, and vegetation. This is critical for creating accurate digital terrain models (DTMs) and other geospatial products. Terrasolid also offers specialized tools for building extraction, which are invaluable in urban planning and development. The software's ability to model ground surfaces accurately is another key feature, making it a valuable asset for infrastructure projects where precise elevation data is necessary.

Functionality and Features

1. **Point Cloud Classification**: Terrasolid provides advanced algorithms for classifying LiDAR point clouds into various categories, such as ground, vegetation, buildings, and other objects. This classification is essential for generating accurate terrain models and for further analysis in urban planning and environmental monitoring. The software's classification tools are known for their precision, which is crucial when dealing with large and complex datasets.

2. **Building Extraction**: One of Terrasolid's standout features is its ability to extract building structures from LiDAR data accurately. This feature is essential for urban planning, where detailed 3D models of buildings are required for simulations, planning, and management. The software's tools allow for the automatic extraction of building footprints and roof structures, which can then be used in various architectural and planning applications.

3. **Ground Surface Modeling**: Terrasolid excels in ground surface modeling, providing users with the tools needed to create detailed and accurate Digital Terrain Models (DTMs). These models are critical in infrastructure management, flood risk assessment, and any project where understanding the terrain is crucial. The software's ability to handle large volumes of data ensures that even complex terrains can be modeled with high precision.

4. **Vegetation Analysis**: The software includes robust tools for analyzing vegetation from LiDAR data, which is particularly useful in environmental monitoring, forestry management, and agriculture.

Terrasolid can differentiate between different types of vegetation and provide detailed canopy height models, biomass estimates, and other important metrics that are essential for managing natural resources.

Evaluation of Functionality and Usefulness

Advantages:

- **Industry Standard**: Terrasolid is widely recognized as the industry standard for LiDAR data processing, particularly in applications requiring high accuracy and detailed outputs. Its extensive feature set and reliability make it a trusted tool among professionals in various industries.

- **Extensive Feature Set**: The software offers a comprehensive range of tools that cover all aspects of LiDAR data processing, from classification and modeling to extraction and analysis. This makes it a versatile solution capable of handling diverse projects with varying levels of complexity.

- **High Accuracy**: Terrasolid's tools are known for their precision, particularly in point cloud classification and ground surface modeling. This high level of accuracy is essential for projects where even minor errors can have significant consequences, such as in urban infrastructure management and environmental monitoring.

- **Strong Community Support**: Terrasolid benefits from a strong user community and extensive documentation, which can be invaluable for troubleshooting, learning new techniques, and staying updated with the latest features and best practices.

Limitations:

- **Complex Interface**: While powerful, Terrasolid's interface can be complex and may present a steep learning curve for new users or those unfamiliar with LiDAR data processing. Mastering the software often requires significant training and experience, which can be a barrier for smaller teams or individuals.

- **Resource Intensive**: Due to its extensive capabilities and the large datasets it often handles, Terrasolid can be resource-intensive, requiring high-performance hardware to run efficiently. This may necessitate additional investment in computing resources, particularly for large-scale projects.

Terrasolid stands out as a powerful and versatile software solution for LiDAR data processing, offering a comprehensive range of tools that meet the needs of professionals across various industries. Its precision in point cloud classification, building extraction, and ground surface modeling makes it an indispensable tool for urban planning, infrastructure management, and environmental monitoring. While the software's complex interface may require a steep learning curve, the accuracy and detail it provides more than justify the investment in time and training. For professionals seeking a reliable and industry-standard tool for handling large and complex LiDAR datasets, Terrasolid remains a top choice.

Global Mapper with LiDAR Module
Overview

Global Mapper's LiDAR Module is a powerful add-on to the widely used Global Mapper GIS software, designed to enhance its capabilities for processing and analyzing LiDAR data. Known for its user-friendly interface and extensive support for a wide range of file formats, the LiDAR Module provides a versatile and accessible solution for professionals who need to work with LiDAR data within a broader GIS context. While it may not offer the same level of specialization as some dedicated LiDAR processing software, its integration with other GIS tools and cost-effectiveness make it a popular choice for many users.

Description of Global Mapper with LiDAR Module

The LiDAR Module for Global Mapper extends the software's already robust geospatial capabilities by adding specialized tools for working with LiDAR data. It is designed to streamline the process of managing, visualizing, and analyzing point clouds, making it easier for users to incorporate LiDAR data into their existing GIS workflows. The module is particularly well-suited for users who need to integrate LiDAR data with other types of geospatial information, such as satellite imagery, vector data, and elevation models.

Global Mapper with the LiDAR Module supports a wide range of LiDAR data formats, allowing users to import, process, and export LiDAR data with ease. The module's automatic point cloud classification and feature extraction tools enable users to quickly categorize and analyze LiDAR data. At the same time, its terrain analysis and 3D visualization capabilities provide detailed insights into the structure of the landscape. This combination of features makes Global Mapper with the LiDAR Module a flexible and powerful tool for applications in surveying, forestry, environmental monitoring, and urban planning.

Functionality and Features

1. **Automatic Point Cloud Classification**: The LiDAR Module includes advanced tools for automatically classifying point clouds into different categories, such as ground, vegetation, and buildings. This automated classification streamlines the data processing workflow, allowing users to quickly generate classified datasets that are ready for further analysis. The classification algorithms are customizable, enabling users to fine-tune the process to meet the specific needs of their projects.

2. **Feature Extraction**: The software offers powerful feature extraction capabilities, allowing users to identify and isolate specific features from LiDAR data, such as building footprints, powerlines, and vegetation. This is particularly useful for urban planning and infrastructure management applications, where accurate feature extraction is essential for creating detailed models and conducting analyses.

3. **Terrain Analysis**: Global Mapper with the LiDAR Module provides comprehensive tools for terrain analysis, including the creation of Digital Elevation Models (DEMs), contour generation, and slope and aspect analysis. These tools are critical for understanding an area's topography, which is important for land use planning, flood risk assessment, and infrastructure development. The software's ability to handle large datasets ensures that even extensive terrain can be analyzed effectively.

4. **3D Visualization**: The LiDAR Module enhances Global Mapper's 3D visualization capabilities, allowing users to create detailed 3D models of the landscape and visualize LiDAR data in a three-dimensional space. This feature is invaluable for exploring the structure of the terrain and for

presenting data in a visually intuitive format, which is useful for communicating findings to stakeholders or for further analysis.

Evaluation of Functionality and Usefulness

Advantages:

- **Versatile Integration with GIS Data**: One of Global Mapper's key strengths with the LiDAR Module is its seamless integration with other types of GIS data. This makes it an excellent choice for users who need to combine LiDAR data with other geospatial information, enabling comprehensive analysis and modeling within a single platform.

- **User-Friendly Interface**: The software is known for its intuitive interface, which lowers the learning curve for new users. This user-friendliness, combined with the software's powerful capabilities, makes it accessible to a wide range of professionals, including those who may not specialize in LiDAR data processing.

- **Cost-Effective**: Compared to other specialized LiDAR processing software, Global Mapper with the LiDAR Module is relatively cost-effective, making it an attractive option for smaller firms or organizations with budget constraints.

Limitations:

- **Limited Specialization**: While the LiDAR Module adds significant functionality to Global Mapper, it is not as specialized as some dedicated LiDAR processing software. This can limit its advanced capabilities, particularly for users who require highly specialized tools for complex LiDAR data analysis or for applications that demand the highest levels of precision and detail.

- **Performance with Large Datasets**: Although the software can handle large datasets, it may not perform as efficiently as some more specialized LiDAR processing tools when dealing with extremely high-density point clouds or extensive geographical areas. Users with such needs might require more powerful, dedicated LiDAR processing software.

Global Mapper with the LiDAR Module is a versatile and user-friendly solution for professionals who need to process and analyze LiDAR data within a broader GIS framework. Its ability to automatically classify point clouds, extract features, and perform terrain analysis makes it a powerful tool for a wide range of applications, including surveying, forestry, environmental monitoring, and urban planning. While it may not offer the same level of specialization as some dedicated LiDAR software, its integration with other GIS tools and cost-effectiveness make it an excellent choice for many users, particularly those who need a comprehensive geospatial solution that includes robust LiDAR processing capabilities.

LAStools
Overview

LAStools is a highly efficient suite of command-line tools designed for rapid processing of LiDAR data. It is renowned for its speed and ability to handle large datasets, making it a popular choice among professionals who need to process vast amounts of LiDAR data quickly and efficiently. Despite its command-line interface, which may be less intuitive for users accustomed to graphical user interfaces

(GUIs), LAStools remains a cost-effective and powerful solution for LiDAR data management and processing.

Description of LAStools

LAStools offers a comprehensive set of tools for processing LiDAR data, focusing on tasks such as point cloud compression, filtering, classification, tile creation, and data conversion. The tools are highly optimized for performance, allowing users to process large datasets much faster than many other LiDAR processing software solutions. LAStools is particularly well-suited for scenarios where time and efficiency are critical, such as large-scale topographic surveys, forestry management, and infrastructure monitoring.

The suite is designed to be lightweight and require minimal resources, which makes it ideal for use on a wide range of hardware setups, from powerful workstations to more modest computing environments. LAStools is also known for its versatility, as it supports a variety of LiDAR formats and can be easily integrated into custom workflows through scripting and automation.

Functionality and Features

1. **Point Cloud Compression**: LAStools includes tools like LASzip, which is widely recognized for compressing LiDAR point clouds without losing data. This feature is essential for managing large datasets, making storage more efficient, and reducing the time required for data transfer.

2. **Filtering and Classification**: The suite provides robust tools for filtering and classifying point clouds based on various criteria, such as elevation, intensity, and return number. These capabilities are crucial for preparing data for further analysis, such as generating Digital Elevation Models (DEMs) or conducting vegetation analysis.

3. **Tile Creation**: LAStools can efficiently divide large LiDAR datasets into smaller, manageable tiles, which is particularly useful for large projects or when working with limited computing resources. Tiling also facilitates parallel processing and can significantly speed up the overall workflow.

4. **Data Conversion**: The tools in LAStools support a wide range of LiDAR formats, enabling seamless data conversion between different formats. This flexibility is essential for users who need to integrate LiDAR data from various sources or who work within heterogeneous environments.

Evaluation of Functionality and Usefulness

Advantages:

- **Fast Processing**: LAStools is highly optimized for speed, making it one of the fastest tools available for processing large LiDAR datasets. This is a significant advantage for professionals who need to process data quickly, especially in time-sensitive projects.

- **Lightweight and Efficient**: The suite's lightweight nature allows it to run efficiently on a wide range of hardware, making it accessible to users with varying levels of computing power.

- **Cost-Effective**: LAStools is a cost-effective solution, offering powerful LiDAR processing capabilities without the high costs associated with some other software suites.

- **Extensive Format Support**: The tools support a wide range of LiDAR data formats, ensuring compatibility across different platforms and enabling easy integration into various workflows.

Limitations:

- **Command-Line Interface**: The command-line interface can be a barrier for users who are not comfortable with scripting or who prefer a more visual, GUI-based approach to software. This may require additional training or adjustment for some users.

- **Limited Visualization Tools**: While LAStools excels in data processing, it lacks the advanced visualization capabilities found in some other LiDAR software, which may necessitate the use of additional tools for tasks like 3D modeling or detailed analysis.

LAStools is an incredibly efficient and powerful suite of tools for processing LiDAR data, particularly well-suited for professionals who require fast processing of large datasets. Its command-line interface, while potentially challenging for some users, allows for flexible integration into automated workflows, making it a valuable asset for large-scale projects. Despite its limitations in visualization and ease of use, LAStools remains a top choice for those seeking a cost-effective and highly efficient LiDAR processing solution.

LP360
Overview

LP360, developed by QCoherent Software, is a powerful LiDAR software suite designed to work seamlessly within the ArcGIS environment and other standalone applications. It provides a robust set of tools for extracting features from point clouds, performing volumetric analysis, and visualizing 3D data, making it an essential tool for GIS professionals who need to incorporate LiDAR data into their geospatial workflows.

Description of LP360

LP360 enhances ArcGIS's capabilities by adding specialized tools for working with LiDAR data directly within the GIS platform. This integration allows users to perform advanced analyses and feature extraction without leaving the familiar ArcGIS environment, streamlining the workflow and improving efficiency. LP360 is particularly valuable for users who need to manage and analyze LiDAR data alongside other geospatial datasets, such as vector data, satellite imagery, and raster maps.

The software offers advanced tools for point cloud classification, feature extraction, and 3D visualization, enabling users to create detailed models and perform complex analyses. LP360's ability to handle large LiDAR datasets and its powerful visualization tools make it a versatile solution for applications in urban planning, forestry, environmental monitoring, and infrastructure management.

Functionality and Features

1. **3D Visualization**: LP360 provides advanced 3D visualization tools that allow users to view and interact with LiDAR point clouds in a three-dimensional space. This capability is essential for understanding the spatial relationships within the data and for performing detailed analyses of terrain, structures, and vegetation.

2. **Point Cloud Classification**: The software includes robust tools for classifying point clouds into different categories, such as ground, vegetation, and buildings. This classification is critical for generating accurate Digital Elevation Models (DEMs) and for conducting further analyses, such as flood modeling or vegetation management.

3. **Feature Extraction**: LP360 excels in feature extraction, enabling users to identify and isolate specific features from LiDAR data, such as powerlines, building footprints, and tree canopies. These tools are particularly valuable for urban planning and infrastructure management, where accurate feature extraction is essential for creating detailed models and simulations.

4. **Integration with GIS**: One of LP360's key strengths is its seamless integration with ArcGIS, allowing users to incorporate LiDAR data into their broader GIS workflows. This integration is particularly beneficial for GIS professionals who need to manage and analyze LiDAR data alongside other geospatial datasets, improving the overall efficiency and effectiveness of their projects.

Evaluation of Functionality and Usefulness

Advantages:

- **Seamless Integration with ArcGIS**: LP360's integration with ArcGIS makes it an ideal tool for GIS professionals who need to incorporate LiDAR data into their geospatial workflows. This integration streamlines the workflow and allows for more efficient data management and analysis.

- **Robust Feature Extraction Tools**: The software's advanced feature extraction capabilities make it highly effective for applications in urban planning, infrastructure management, and environmental monitoring.

- **3D Visualization**: LP360's powerful 3D visualization tools provide valuable insights into the structure and relationships within LiDAR data, making it easier to analyze and interpret complex datasets.

Limitations:

- **Tied to the ArcGIS Environment**: While LP360's integration with ArcGIS is a significant advantage for GIS professionals, it may be a limitation for users who do not use ArcGIS or who require a more flexible, standalone solution.

- **Learning Curve for Non-GIS Users**: The software is designed with GIS professionals in mind, which may present a learning curve for users who are less familiar with GIS platforms or who require a more straightforward LiDAR processing tool.

LP360 is a powerful and versatile LiDAR software suite that excels in feature extraction, point cloud classification, and 3D visualization. Its seamless integration with ArcGIS makes it an essential tool for GIS professionals who need to incorporate LiDAR data into their geospatial workflows. While it may be less flexible for non-GIS users, its robust capabilities and integration make it a valuable asset for a wide range of applications, including urban planning, infrastructure management, and environmental monitoring.

Trimble Business Center (TBC)
Overview

Trimble Business Center (TBC) is a comprehensive office software suite that supports the processing of geospatial data from various sources, including LiDAR. Widely used in surveying, civil engineering, and construction, TBC provides powerful tools for LiDAR data processing, point cloud management, and surface modeling, making it an integral part of the workflow for professionals who rely on Trimble hardware and software solutions.

Description of Trimble Business Center

TBC is designed to be a central hub for managing and processing geospatial data, offering a wide range of tools for analyzing data collected from LiDAR, GNSS, and other surveying technologies. The software is particularly well-suited for users within the Trimble ecosystem, providing seamless integration with Trimble hardware such as scanners, GNSS receivers, and total stations. This integration allows for a smooth transition from data collection to processing, analysis, and reporting.

TBC's LiDAR processing capabilities are extensive, enabling users to manage large point clouds, classify data, and generate accurate surface models. The software also offers advanced tools for creating Digital Terrain Models (DTMs), conducting volumetric analysis, and performing quality control on LiDAR data. These features are particularly valuable for civil engineering projects, where precise measurements and detailed surface models are essential for planning, design, and construction.

Functionality and Features

1. **LiDAR Data Processing**: TBC provides comprehensive tools for processing LiDAR data, including point cloud management, classification, and surface modeling. The software can handle large datasets, making it suitable for complex projects that require detailed analysis and accurate modeling.

2. **Point Cloud Management**: The software includes powerful tools for filtering, classifying, and editing LiDAR data. These tools are essential for creating clean and accurate datasets that can be used for further analysis and modeling.

3. **Surface Modeling**: TBC excels in surface modeling, providing users with the tools needed to create detailed Digital Terrain Models (DTMs) and Digital Surface Models (DSMs). These models are critical for applications in civil engineering, construction, and land development, where accurate terrain data is necessary for planning and design.

4. **Integration with Trimble Hardware**: One key advantage of TBC is its seamless integration with Trimble hardware, allowing for a smooth workflow from data collection to processing. This integration ensures compatibility and optimizes the accuracy and efficiency of the overall geospatial data management process.

Evaluation of Functionality and Usefulness

Advantages:

- **Integrated Workflow with Trimble Hardware**: TBC's integration with Trimble hardware provides a seamless workflow for users within the Trimble ecosystem, improving efficiency and ensuring accuracy across all stages of the data management process.

- **Powerful Tools for Civil Engineering Projects**: The software's advanced surface modeling and LiDAR data processing capabilities make it an ideal tool for civil engineering, surveying, and construction projects.

- **Comprehensive Geospatial Data Management**: TBC serves as a central hub for managing a wide range of geospatial data, offering a comprehensive suite of tools for analysis, modeling, and reporting.

Limitations:

- **Best Suited for Trimble Ecosystem Users**: While TBC offers powerful tools, it is best suited for users who are already within the Trimble ecosystem. Those who use different hardware or who require a more versatile solution may find it less accommodating.

- **Cost**: TBC is a high-end software solution, and its advanced capabilities and integration with Trimble hardware come at a premium price, which may be a barrier for smaller firms or users with limited budgets.

Trimble Business Center (TBC) is a comprehensive and powerful software solution for managing and processing geospatial data, particularly for users within the Trimble ecosystem. Its robust tools for LiDAR data processing, surface modeling, and point cloud management make it an essential tool for civil engineering, surveying, and construction projects. While it is best suited for those who already use Trimble hardware, its advanced capabilities and integrated workflow make it a top choice for professionals seeking a reliable and accurate geospatial data management solution.

Merrick MARS
Overview

Merrick Advanced Remote Sensing (MARS) is a specialized software suite designed for the processing, visualization, and analysis of high-density LiDAR data. It is widely used in applications such as topographic mapping, forestry, and infrastructure management, where accuracy and detailed analysis are crucial. MARS is known for its ability to handle large, complex datasets and for providing industry-specific tools that cater to the unique needs of different sectors.

Description of Merrick MARS

MARS is designed to process large volumes of LiDAR data efficiently, providing users with the tools needed to create accurate terrain models, conduct volumetric calculations, and perform detailed analyses of vegetation and infrastructure. The software is particularly well-suited for high-density point cloud processing, making it an excellent choice for projects that require detailed and precise data, such as forestry management, urban planning, and flood risk assessment.

MARS offers a range of specialized tools for different industries, including forestry analysis, terrain modeling, and infrastructure management. These tools allow users to perform specific tasks such as canopy height modeling, volumetric analysis, and slope stability assessment. The software's ability to process and visualize large datasets in 3D is also a significant advantage, enabling users to explore and analyze complex data in a visually intuitive format.

Functionality and Features

1. **High-Density Point Cloud Processing**: MARS is optimized for processing high-density LiDAR point clouds, allowing users to manage and analyze large datasets with precision. This capability is

essential for projects that require detailed data, such as topographic mapping and infrastructure management.

2. **Terrain Modeling**: The software includes advanced tools for creating Digital Elevation Models (DEMs), Digital Surface Models (DSMs), and contour maps. These terrain models are critical for applications in land development, flood risk assessment, and civil engineering, where accurate elevation data is essential.

3. **Forestry Analysis**: MARS offers specialized tools for forestry analysis, including canopy height modeling, biomass estimation, and vegetation classification. These tools are invaluable for managing forest resources, assessing forest health, and conducting environmental impact studies.

4. **Volumetric Calculations**: The software provides powerful tools for performing volumetric calculations, such as cut-and-fill analysis and material stockpile measurement. This capability is instrumental in construction, mining, and land development projects, where precise volume measurements are necessary for planning and budgeting.

Evaluation of Functionality and Usefulness

Advantages:

- **High Accuracy**: MARS is designed to deliver high accuracy in LiDAR data processing, making it ideal for applications where precision is critical, such as topographic mapping, forestry management, and infrastructure analysis.

- **Specialized Tools for Various Industries**: The software's industry-specific tools cater to the unique needs of different sectors, providing users with the functionality required to perform detailed analysis and modeling.

- **Scalable for Large Projects**: MARS is capable of handling large-scale projects with ease, making it suitable for organizations that need to process and analyze vast amounts of LiDAR data efficiently.

Limitations:

- **Complexity**: MARS's advanced features and specialized tools may present a learning curve for new users, particularly those who are not familiar with LiDAR data processing. This complexity can require additional training and time investment to utilize the software's capabilities fully.

- **Cost**: MARS is a high-end software solution, and its advanced capabilities come with a corresponding price. This may be a consideration for smaller organizations or those with limited budgets.

Merrick MARS is a powerful and specialized software suite designed for high-density LiDAR data processing and analysis. Its industry-specific tools for terrain modeling, forestry analysis, and infrastructure management make it an invaluable resource for professionals who require precise and detailed data. While its complexity may require additional training, and its cost may be a consideration for some users, MARS's scalability, accuracy, and specialized capabilities make it an excellent choice for large-scale projects and industry-specific applications.

Conclusion

The landscape of drone image processing software is rich and varied, with each platform offering distinctive tools and capabilities that cater to the diverse needs of users across multiple industries. From the high-precision and feature-rich environments of Global Mapper Pro and Mapware to the user-friendly interfaces of DroneDeploy and Maps Made Easy, these platforms enhance the ability of professionals to execute projects with greater accuracy and efficiency. Additionally, specialized platforms like DJI Terra and Autel Mapper provide integrated solutions that streamline workflows for specific drone models. Meanwhile, Propeller Aero emphasizes accuracy with its unique AeroPoints technology, offering a significant advantage in survey accuracy. By leveraging these advanced software solutions, businesses can harness the full potential of drone technology, driving innovation and providing detailed, reliable data that is crucial for decision-making and strategic planning. As drone technology continues to evolve, the capabilities and applications of these software platforms will undoubtedly expand, further embedding them as essential tools in the toolkit of geospatial and survey professionals worldwide.

Chapter 9: Ancillary Software

What is this ancillary software you speak of, and why do I need it?

The term "ancillary" refers to something that provides support for or is supplementary to a primary function or activity. In this case, there are a few different types of software that can be referred to as ancillary in nature. This is by no means a complete list; in fact, it is likely a very fragmented one. However, there is a need to discuss some of these software packages for what will become Obvious reasons. And others, well, they're just useful.

One of the most essential tools you'll use as a drone pilot, if you aren't already, is airspace access software or websites. Platforms like faadronezone.faa.gov and software such as aloft.ai are key resources. The National Airspace System (NAS), which encompasses all airspace above navigable ground within the United States, is managed by the FAA under the authority granted by Section 307(a) of the Federal Aviation Act of 1958. While other countries have their own laws governing airspace systems, this discussion will focus on U.S. regulations. If you're reading this from outside the United States, I strongly recommend familiarizing yourself with the airspace regulations in your region, as airspace management can vary significantly from one country to another.

Understanding Airspace in the United States
The National Airspace System (NAS)

To understand the need for drone integration into the NAS, it is first paramount to understand the NAS.

The National Airspace System (NAS) is an integrated network of airspace, navigation facilities, and airports that supports the operation of aircraft in the United States. The NAS is designed to provide a safe, efficient, and flexible environment for all users, from commercial airliners to private aircraft, commonly referred to as GA or General Aviation, to all outdoor drone operations. It incorporates advanced technologies, air traffic control (ATC) systems, and comprehensive regulations to manage the diverse activities within U.S. airspace.

The single most critical component of the NAS is the air traffic control system, which ensures that aircraft can travel safely from one point to another without collision. This system involves thousands of controllers, who monitor and guide aircraft through various phases of flight, from takeoff to landing. Additionally, the NAS includes navigation aids, such as GPS and ground-based systems, which provide precise location data to pilots and ATC.

In the United States, airspace is classified into various classes (A, B, C, D, E, and G), each with specific rules and regulations. This classification helps manage the complex environment of aviation, ensuring safety and efficiency. As stated above, the Federal Aviation Administration (FAA) is the regulatory body responsible for overseeing U.S. airspace, implementing guidelines, and maintaining order within this domain.

Class A airspace, for instance, begins at 18,000 feet above mean sea level and extends up to 60,000 feet. This airspace is primarily for high-altitude, high-speed aircraft and requires Instrument Flight Rules (IFR) for navigation. Classes B, C, and D are associated with airspace surrounding towered airports, with Class B covering the busiest, such as Dallas Fort Worth (DFW), Los Angeles International (LAX), and John F. Kennedy International (JFK). Class G, the only uncontrolled airspace, is at the opposite end of the spectrum, typically found in less populated, rural areas. Classes of airspace are covered in detail in the Part 107 study material, and it is critical for every Part 107 pilot to understand airspace and know their limitations when operating within it.

Drone Operations and the NAS

With the rise of unmanned aerial vehicles (UAVs), aircraft we commonly refer to as drones, the NAS has had to adapt to incorporate these new airspace users. The FAA has established specific regulations and guidelines for drone operations to ensure they do not interfere with manned aircraft and maintain safety standards.

Drones are generally restricted to Class G airspace for recreational use and are required to stay below 400 feet above ground level (AGL) to minimize conflict with traditional aircraft. However, commercial drone operations, such as those we have discussed in this text, often require access to controlled airspace (Classes B, C, D, and E). In such cases, operators may be required to obtain authorization through the FAA's Low Altitude Authorization and Notification Capability (LAANC) system, which provides near-real-time approval for many operations in controlled airspace. In instances where a LAANC is not available, it is necessary to submit an approval request directly to the FAA Through the FAADronezone portal.

Furthermore, drones must adhere to specific operational guidelines, such as maintaining a visual line of sight, not exceeding certain speed limits, and avoiding restricted or prohibited areas like military installations and critical infrastructure. As drone technology and applications continue to evolve, the NAS will likely see further integration of UAVs, necessitating ongoing updates to regulations and airspace management strategies to ensure a harmonious coexistence between manned and unmanned aircraft.

FAA DroneZone Website:

The FAA DroneZone is an official platform created by the Federal Aviation Administration (FAA) to manage and streamline the registration, certification, and operations of unmanned aircraft systems. Here are some of the key features and functionalities of the FAA DroneZone website:

Key Features of FAA DroneZone

1. **Registration:**

    o **Individual Registration:** Hobbyists and recreational drone flyers can register their drones, which is required for all drones weighing between 0.55 pounds and 55 pounds.

    o **Part 107 Registration:** For commercial drone operators flying under the FAA's Part 107 regulations, the website provides the necessary forms and procedures to register their drones. All drones operating commercially require registration regardless of weight up to 55lbs. UAS heavier than 55lbsd require separate registration which is not covered in this text.

2. **Remote Pilot Certification:**

    o The site offers resources for obtaining a Remote Pilot Certificate, which is required for commercial drone operations. This includes access to study materials, scheduling for the knowledge test, and instructions for submitting an application.

3. **Operational Waivers and Authorizations:**

- **Waivers:** Operators can apply for waivers to bypass certain Part 107 regulations, such as Flying over moving cars, over people, or beyond the visual line of sight.

- **Authorizations:** Drone pilots can request authorization to fly in controlled airspace near airports through the Low Altitude Authorization and Notification Capability (LAANC) system. If LAANC is not available in or around certain airports, a manual request can be submitted on the website. These requests can take anywhere from 10 days to 90 days, so pilots should plan accordingly.

4. **Reporting and Compliance:**

- The DroneZone platform allows users to report any incidents or accidents involving their drones, which is a requirement for significant events like injuries or property damage.

- Users can also manage and track their compliance with FAA regulations through their account on the website.

5. **Educational Resources:**

- The website provides a wealth of information and resources, including guides, FAQs, and instructional videos, to help drone operators understand and comply with FAA regulations.

Benefits of Using FAA DroneZone

- **Centralized Management:** The platform centralizes all necessary tools and information for drone operators, making it easier to manage registration, certification, and operational compliance.

- **User-Friendly Interface:** The website is designed to be user-friendly, with clear instructions and straightforward processes for various applications and submissions.

- **Centralized Authorizations:** Through the integration with the LAANC system, drone operators can apply for and receive approvals for flight operations in controlled airspace that are not available on the online LAANC system, facilitating more efficient and timely missions.

- **Enhanced Safety:** The FAA DroneZone helps promote safer skies and better accountability for drone operations by ensuring that all drones are registered and operators are certified.

Accessing FAA DroneZone

To use the FAA DroneZone, users need to create an account on the platform. Once registered, they can log in to access all the services provided. The website is accessible at FAA DroneZone, where users can find detailed instructions and resources related to their specific needs.

In summary, the FAA DroneZone website is an essential tool for drone operators in the United States. It provides a centralized hub for registration, certification, compliance, and education. It plays a crucial role in integrating drones into the National Airspace System (NAS) while ensuring safety and regulatory compliance.

Aloft.ai

Aloft.ai is a leading company specializing in providing innovative solutions for drone operations, safety, and airspace management. Formerly known as Kittyhawk, Aloft.ai offers a suite of tools and services designed to enhance the capabilities of both recreational and commercial drone operators. With a focus on making drone flights safer and more efficient, Aloft.ai has established itself as a crucial player in the unmanned aircraft systems (UAS) industry.

Leading Provider of LAANC Authorizations

Aloft.ai is recognized as the number one provider of Low Altitude Authorization and Notification Capability (LAANC) authorizations in the United States. LAANC is a collaboration between the Federal Aviation Administration (FAA) and private industry that automates the application and approval process for drone flights in controlled airspace. This system allows drone operators to receive near-real-time airspace authorizations, significantly streamlining the process and ensuring compliance with regulatory requirements.

Key Features of Aloft.ai's LAANC Services:

1. **Real-Time Approvals:**

   o Through its LAANC integration, Aloft.ai provides immediate authorization for drone flights in controlled airspace. This allows operators to plan and execute missions with minimal delays.

2. **User-Friendly Interface:**

   o The platform offers an intuitive and easy-to-navigate interface, making it accessible for both novice and experienced drone pilots. The streamlined process ensures that users can quickly submit requests and receive approvals.

3. **Comprehensive Airspace Data:**

   o Aloft.ai integrates extensive airspace data, including information on no-fly zones, temporary flight restrictions (TFRs), and other relevant airspace notifications. This helps operators stay informed and make safer flight decisions.

4. **Compliance and Safety:**

   o By facilitating LAANC authorizations, Aloft.ai ensures that drone operations comply with FAA regulations, promoting safer skies and reducing the risk of incidents in controlled airspace.

5. **Mobile and Web Access:**

   o Aloft.ai's LAANC services are accessible via its mobile app AIRCONTROL and its web platform, www.aloft.ai, providing flexibility and convenience for drone operators to manage their flight plans on the go.

Impact on the Drone Industry

Aloft.ai's dominance in providing LAANC authorizations has had a significant impact on the drone industry. By simplifying the process of obtaining airspace approvals, Aloft.ai has enabled more widespread and efficient use of drones for various applications, including aerial photography, agriculture, infrastructure inspection, and emergency response. The company's commitment to safety, compliance, and innovation has made it a trusted partner for drone operators across the United States.

Aloft.ai stands out as the premier provider of LAANC authorizations in the U.S., playing a pivotal role in integrating drones into the National Airspace System (NAS). Through its advanced technology and user-centric approach, Aloft.ai has transformed the way drone operators navigate controlled airspace, ensuring safe, efficient, and compliant operations. As the drone industry continues to grow, Aloft.ai remains at the forefront, driving innovation and setting new standards for airspace management and drone safety.

Chapter 10: The Missions

Understanding Photogrammetry, Mapping, and Survey Missions

Introduction

Similarities

Before delving into the differences, it's important to acknowledge the common ground between Photogrammetry, Mapping, and Survey missions. Each of these missions involves using drones equipped with cameras or sensors to capture data from above. They rely on overlapping images or sensor readings to create detailed representations of the area of interest. Additionally, all three require precise flight planning, data processing, and analysis to produce usable outputs. However, the nature of these outputs and the specific techniques employed vary significantly.

Photogrammetry Missions

Definition and Purpose:

Photogrammetry is the science of making measurements from photographs. In a photogrammetry mission, the primary objective is usually to create detailed 3D models and maps by capturing a series of high-resolution images from multiple angles. These images are then processed using specialized software to generate orthomosaic maps, digital elevation models (DEMs), and 3D models.

Key Requirements:

High Overlap: Photogrammetry requires a high degree of overlap between images, typically around 70-80% front overlap and 60-80% side overlap.

Multiple Angles: To create accurate 3D models, images are captured from various angles, including oblique (angled) and Nadir (straight down) shots.

Post-Processing: Significant post-processing is required to stitch images together and create detailed 3D models and maps.

Applications: Common applications include construction site monitoring, cultural heritage documentation, and creating virtual environments.

Use Cases for Photogrammetry
Construction Site Monitoring

Description: Photogrammetry is used to create detailed 3D models of construction sites, allowing project managers to monitor progress, plan logistics, and ensure that the construction is proceeding according to design specifications. These models help in tracking changes over time and identifying potential issues early.

Cultural Heritage Documentation

Description: In the preservation of historical sites and artifacts, photogrammetry is employed to create accurate 3D models. This allows for detailed analysis and preservation without physical interference. These models can be used for virtual tours, restoration planning, and archival purposes.

Mining and Quarrying

Description: Photogrammetry is utilized in mining operations to create 3D models of mining sites. These models assist in monitoring the progress of extraction, calculating volumes of extracted materials, and ensuring that operations are carried out safely and efficiently.

Forestry and Environmental Management

Description: In forestry, photogrammetry helps in creating detailed maps and 3D models of forested areas. These models are used for inventory management, monitoring deforestation, and planning reforestation efforts. In environmental management, it aids in tracking changes in landscapes and ecosystems.

Urban Planning and Development

Description: Urban planners use photogrammetry to create detailed models of cities and towns. These models help in planning new developments, assessing the impact of proposed projects, and visualizing changes to the urban landscape. They also assist in creating zoning plans and infrastructure development.

Agricultural Management

Description: In agriculture, photogrammetry is used to create high-resolution maps of farmland. These maps help in precision farming by providing data on crop health, soil conditions, and irrigation needs. This technology enables farmers to optimize resource use and increase crop yields.

Disaster Response and Management

Description: After natural disasters such as earthquakes, floods, or hurricanes, photogrammetry is used to create detailed maps of affected areas. These maps help in assessing damage, planning relief efforts, and coordinating response activities. They provide crucial information for rebuilding and recovery efforts.

Archaeological Surveys

Description: In archaeology, photogrammetry is employed to create detailed 3D models of excavation sites and artifacts. These models provide accurate records of archaeological findings, help in the analysis of historical contexts, and support the preservation of fragile sites and objects.

Infrastructure Inspection and Maintenance

Description: Photogrammetry is used to inspect and monitor infrastructure such as bridges, roads, and buildings. By creating detailed 3D models, engineers can detect structural issues, plan maintenance, and ensure the safety and integrity of these structures. Using drones allows for regular, safer, non-invasive inspections.

Film and Game Development

Description: In the entertainment industry, photogrammetry is used to create realistic 3D models for films and video games. By capturing real-world environments and objects, developers can create immersive virtual worlds. This technology enhances the visual quality and realism of digital media.

These varied applications of photogrammetry showcase its versatility and importance across multiple fields, from construction and urban planning to environmental management and entertainment. Each use case benefits from the precise and detailed data that photogrammetry provides, enabling better decision-making and outcomes.

Mapping Missions

Definition and Purpose:
Mapping missions focus on creating accurate, georeferenced maps of an area. Unlike photogrammetry, which emphasizes 3D modeling, mapping missions are primarily concerned with creating 2D representations of the terrain. These missions often involve the use of orthophotos, which are aerial photographs corrected for lens distortion, camera tilt, and topographic relief.

Key Features:

Orthophotos: The primary output of a mapping mission is an orthophoto, which provides a true-to-scale map of the area.

Lower Overlap: Mapping missions typically require less image overlap compared to photogrammetry missions, usually around 60% front overlap and 30% side overlap.

Georeferencing: Accurate georeferencing is crucial, often involving the use of ground control points (GCPs) to ensure spatial accuracy.

Applications: Mapping missions are commonly used in urban planning, agriculture, and environmental monitoring.

10 Uses for Drone Mapping

Agricultural Monitoring

Description: Drone mapping in agriculture provides high-resolution maps of fields, allowing farmers to monitor crop health, assess soil conditions, and manage irrigation systems efficiently. These maps help in detecting pest infestations, nutrient deficiencies, and other issues, enabling precision farming techniques that optimize yield and reduce resource use.

Construction Site Management

Description: Drones are used to create detailed maps of construction sites, offering real-time progress updates and accurate measurements. These maps help project managers track the development stages, ensure compliance with design plans, identify potential issues, and improve site safety by monitoring hazardous areas.

Disaster Response and Management

Description: After natural disasters such as earthquakes, floods, or hurricanes, drone mapping provides critical information for assessing damage and planning relief efforts. High-resolution maps help emergency responders locate affected areas, evaluate the extent of destruction, and coordinate rescue operations efficiently.

Environmental Conservation

Description: Drones are used to map and monitor natural habitats, track wildlife populations, and assess the health of ecosystems. These maps support conservation efforts by providing data on deforestation, illegal logging, habitat fragmentation, and the impacts of climate change, helping to protect endangered species and manage natural resources.

Urban Planning and Development

Description: Drone mapping aids urban planners in creating accurate and up-to-date maps of cities and towns. These maps help in visualizing current infrastructure, land use, and development patterns, assisting in designing new projects, improving zoning regulations, and planning for sustainable urban growth.

Infrastructure Inspection

Description: Drones equipped with mapping technology are used to inspect critical infrastructure such as bridges, roads, and power lines. Detailed maps allow engineers to detect structural issues, plan maintenance, and ensure the integrity and safety of these structures without the need for manual inspections.

Mining and Quarrying

Description: In mining operations, drone mapping provides detailed 3D maps of mining sites. These maps help in monitoring extraction progress, calculating volumes of extracted materials, and planning future operations. They also assist in ensuring compliance with safety and environmental regulations.

Forestry Management

Description: Drones are used to map forested areas, providing data on tree health, species distribution, and biomass. These maps support sustainable forest management practices by helping foresters plan logging activities, monitor reforestation efforts, and assess the impact of pests and diseases on forest health.

Coastal and Marine Surveys

Description: Drone mapping is employed to survey coastal and marine environments, providing data on shoreline changes, coral reef health, and underwater habitats. These maps help in managing coastal resources, planning conservation efforts, and assessing the impacts of human activities and natural events on marine ecosystems.

Real Estate and Property Management

Description: In the real estate industry, drones are used to create detailed maps and 3D models of properties. These maps help in marketing properties by providing potential buyers with comprehensive views of the land and structures. They also assist property managers in planning developments, managing large estates, and ensuring property boundaries are accurately defined.

Each of these applications highlights the versatility and value of drone mapping across various industries. Drones provide precise, real-time data that enhance decision-making, improve efficiency, and support sustainable practices in agriculture, construction, environmental conservation, urban planning, and beyond.

Survey Missions

Definition and Purpose:
Survey missions are conducted to gather precise geospatial data for specific purposes, often involving land surveying, construction, and infrastructure development. These missions are characterized by their emphasis on accuracy and precision. Survey missions can use various sensors, including cameras, LiDAR, and GNSS receivers, to collect data. As previously discussed, it is important to note that surveys can only be carried out by licensed surveyors, and in some cases civil engineers. Do not assume that you can buy a drone, even an expensive one equipped with RTK, a base station, and ground control points and that you can start conducting surveys. Always check with local laws to verify that you are in compliance before attempting to perform or sell these services.

Key Features:

High Precision: Survey missions require a high level of precision, often down to centimeter-level accuracy.

Specialized Equipment: In addition to cameras, survey missions may use LiDAR sensors and GNSS receivers for precise measurements.

Regulatory Compliance: Survey missions must adhere to strict regulatory standards and guidelines, often involving licensed surveyors.

Applications: Common applications include boundary surveys, topographic surveys, and construction site layout.

Differences in Workflow

Planning:

Photogrammetry: Requires detailed flight planning to ensure high overlap and capture images from multiple angles. The flight path is often complex and may include multiple passes over the same area.

2D Mapping: Flight planning is simpler, focusing on achieving sufficient coverage and overlap to create accurate orthomosaics.

Survey: Involves precise planning to ensure coverage of specific areas and the use of ground control points for accuracy.

Data Collection:

Photogrammetry: Data collection involves capturing a large number of high-resolution images from different angles.

Mapping: Focuses on capturing orthophotos, with less emphasis on multiple angles.

Survey: May involve a combination of photographic, LiDAR, and GNSS data collection.

Post-Processing:

Photogrammetry: Extensive post-processing to create 3D models and orthomosaic maps.

Mapping: Processing orthophotos to ensure accurate georeferencing and stitching.

Survey: Precision processing to ensure data accuracy, often involving specialized software and equipment.

Uses for Drone Surveying

Topographic Surveys

Description: Drone surveys are extensively used to create detailed topographic maps that show the contours and features of the land. These surveys are essential for planning construction projects, land development, and landscape design. The high-resolution data collected helps engineers and architects design projects that are well-suited to the terrain.

Land Boundary Surveys

Description: Drones can quickly and accurately survey land boundaries, providing precise measurements and detailed maps that are crucial for property ownership and real estate transactions. These surveys ensure that property lines are accurately defined and documented, helping to resolve disputes and avoid legal issues.

Construction Site Surveys

Description: Drone surveys provide up-to-date information on construction sites, allowing project managers to monitor progress, verify the accuracy of work, and manage logistics. The high-resolution images and 3D models generated help in identifying potential issues, ensuring that the project stays on track and within budget.

Infrastructure Inspection

Description: Drones are used to survey and inspect infrastructure such as bridges, roads, and power lines. These surveys provide detailed visual and thermal data that can identify structural issues, wear and tear, and other maintenance needs. This information helps in planning repairs and ensuring the safety and longevity of infrastructure assets.

Environmental Impact Assessments

Description: Drone surveys are employed to assess the environmental impact of construction projects, mining operations, and other activities. They provide detailed data on vegetation cover, water bodies, and wildlife habitats, helping environmentalists and planners minimize ecological disruption and ensure compliance with environmental regulations.

Agricultural Surveys

Description: In agriculture, drones are used to survey large fields and provide data on crop health, soil conditions, and irrigation needs. These surveys help farmers implement precision agriculture techniques, optimizing the use of water, fertilizers, and pesticides, which leads to increased yields and reduced environmental impact.

Mining Surveys

Description: Drone surveys in mining provide detailed maps of mining sites, helping to monitor extraction activities, calculate volumes of extracted materials, and plan future operations. These surveys also ensure compliance with safety and environmental regulations by providing accurate and up-to-date data.

Flood Risk Management

Description: Drones are used to survey areas prone to flooding, providing detailed topographic data that helps in modeling flood scenarios and planning mitigation measures. These surveys are crucial for designing flood defenses, improving drainage systems, and developing emergency response plans.

Archaeological Surveys

Description: Drones offer a non-invasive way to survey archaeological sites, providing detailed maps and 3D models of excavation areas. These surveys help archaeologists document and analyze sites, plan excavations, and preserve cultural heritage by minimizing physical disturbance to the site.

Forest Management

Description: Drone surveys are used to monitor forests, providing data on tree health, species distribution, and biomass. These surveys support sustainable forest management practices by helping foresters plan logging activities, monitor reforestation efforts, and assess the impact of pests and diseases on forest ecosystems.

Each of these uses demonstrates the importance of drone surveying in various fields. The precise and detailed data collected by drones enhance decision-making, improve efficiency, and support sustainable practices in construction, agriculture, environmental management, and beyond.

Choosing the Right Mission

Selecting the appropriate mission depends on the specific needs and objectives of the project. For projects requiring detailed 3D models, oblique photogrammetry is the best choice. Mapping missions are ideal for creating accurate 2D maps for urban planning and environmental monitoring. Survey missions are essential for projects requiring high precision, such as land surveys and construction site layouts.

Understanding the differences between Photogrammetry, Mapping, and Survey missions is crucial for drone operators and their clients. Each type of mission serves a unique purpose and requires specific techniques and equipment. By choosing the right mission for the job, operators can ensure they deliver accurate and valuable data to meet their clients' needs. This knowledge empowers operators to optimize their drone services, ensuring successful outcomes and satisfied clients.

LiDAR Missions
Definition and Purpose

LiDAR (Light Detection and Ranging) uses pulses of laser light to measure variable distances to the Earth's surface. By emitting thousands—or even millions—of laser pulses per second and measuring the time it takes for each pulse to return, LiDAR sensors create highly detailed point clouds of the terrain or structures below. These point clouds can be processed to form precise 3D models and elevation maps (such as digital terrain models, DTMs, and digital surface models, DSMs).

LiDAR missions are particularly valuable when extreme accuracy is required or when certain features cannot be easily captured by photogrammetry (e.g., under dense vegetation). As an active sensor, LiDAR can operate in varying light conditions, including low light and nighttime.

Key Features

Active Sensing: LiDAR does not rely on ambient light; it emits its own laser pulses, allowing for data collection in low-light or night conditions.

High Accuracy: Provides centimeter-level precision in elevation data, making it ideal for engineering and survey-grade projects.

Vegetation Penetration: Near-infrared LiDAR can often penetrate through tree canopies, enabling the creation of bare-earth models even in forested areas.

High Point Density: Can collect millions of points per second, resulting in highly detailed 3D data.

Complex Post-Processing: Requires specialized software to handle point cloud classification, noise filtering, and model generation.

Applications: Commonly used for topographic surveys, corridor mapping (roads, railways, utility lines), flood modeling, forestry management, and archaeology.

Uses for LiDAR Missions

Topographic Mapping and DEM Generation
Description: LiDAR data is used to create highly accurate Digital Elevation Models (DEMs). These DEMs are essential for infrastructure planning, earthworks, and large-scale engineering projects, providing precise terrain contours and slope information.

Forestry and Vegetation Analysis
Description: Because LiDAR can penetrate sparse and moderate canopy cover, it allows foresters to see the ground beneath the trees. This helps in estimating timber volume, monitoring forest health, and planning sustainable logging or reforestation efforts.

Power Line and Utility Corridor Inspections
Description: LiDAR missions can map power lines and utility corridors with great detail. This data helps identify vegetation encroachment, conductor sag, and infrastructure defects, improving safety and reducing maintenance costs.

Flood Risk Assessment and Watershed Management
Description: High-resolution elevation data from LiDAR is invaluable for modeling water flow and predicting flood scenarios. It helps urban planners and environmental agencies design better drainage systems and flood mitigation strategies.

Transportation Infrastructure
Description: LiDAR mapping of roads, bridges, and railways aids in design, construction, and maintenance. Accurate 3D models help engineers plan expansions, detect surface deformities, and ensure compliance with safety standards.

Archaeological Site Mapping
Description: LiDAR can reveal hidden structures, such as ancient roads, building foundations, and earthworks, even under dense vegetation. Archaeologists use these detailed elevation models to plan excavations and preserve cultural heritage sites.

Coastal Zone Management
Description: LiDAR is used to monitor coastline erosion, sea-level rise, and changes in beach profiles. This data informs coastal conservation efforts, infrastructure placement, and disaster preparedness.

Urban Mapping and Planning
Description: In densely built environments, LiDAR quickly produces precise 3D models of buildings, streets, and utilities. Urban planners use this data for zoning, infrastructure upgrades, and assessing the impact of proposed construction.

Mining Operations and Stockpile Analysis
Description: LiDAR can accurately measure volumes of stockpiles in mining and quarrying operations, providing real-time data on resource extraction. This improves inventory management and ensures compliance with safety regulations.

Pipeline and Oil & Gas Industry
Description: In pipeline routing and monitoring, LiDAR data helps detect geohazards, slope instability,

and ground movement. By identifying potential risks early, companies can prevent leaks, spills, and costly accidents.

Multispectral Missions
Definition and Purpose

Multispectral missions involve capturing images across different bands of the electromagnetic spectrum—commonly red, green, blue (RGB), near-infrared (NIR), and red-edge. These sensors gather data on the reflectance of vegetation and other surfaces, providing insights into plant health, soil conditions, and environmental factors that are not visible to the naked eye.

Multispectral data is widely used in precision agriculture, environmental monitoring, and land management. By analyzing reflectance patterns, professionals can identify stress in crops, monitor deforestation, and optimize resource usage.

Key Features

Multiple Spectral Bands: Collects data in discrete wavelengths such as red, green, blue, near-infrared, and red-edge.

Vegetation Health Analysis: Enables the creation of indices like NDVI (Normalized Difference Vegetation Index) to measure plant vigor.

Precision Agriculture: Allows targeted interventions (fertilizers, pesticides, irrigation) based on plant health and soil conditions.

Environmental Monitoring: Facilitates tracking of deforestation, water pollution, and habitat changes over time.

Data Integration: Can be combined with other datasets (e.g., LiDAR or thermal) for comprehensive analysis.

Applications: Widely used in agriculture, forestry, environmental conservation, and resource management.

Uses for Multispectral Missions
Crop Health and Stress Detection
Description: By analyzing reflectance in NIR and red-edge bands, farmers can identify areas of crop stress, pest infestation, or nutrient deficiency. This guides targeted treatments, improving yields and reducing costs.

Precision Fertilization and Irrigation
Description: Multispectral imagery helps determine exactly where and when to apply fertilizers and water, minimizing waste and environmental impact while maximizing crop production.

Disease and Pest Monitoring

Description: Early detection of diseases or pest outbreaks is possible by analyzing changes in plant reflectance. This allows timely intervention and prevents widespread crop loss.

Soil Analysis

Description: Variations in soil composition can be identified through multispectral data. Farmers and land managers can use this information to apply soil amendments only where needed, optimizing cost and productivity.

Reforestation and Afforestation

Description: Multispectral imagery is used to monitor the survival and health of newly planted forests. It enables foresters to evaluate the success of reforestation efforts and take corrective measures if necessary.

Habitat and Wetland Monitoring

Description: Environmental agencies utilize multispectral data to track changes in wetlands, marshes, and other sensitive habitats. Changes in vegetation cover or water quality are detected through shifts in spectral signatures.

Invasive Species Management

Description: Multispectral sensors can detect invasive plant species by identifying distinct spectral patterns, allowing for early removal or containment strategies to protect native ecosystems.

Urban Green Space Analysis

Description: Cities use multispectral data to assess the health of urban vegetation—trees, parks, and gardens. This information supports urban planning aimed at improving air quality and quality of life.

Water Resource Management

Description: By analyzing water bodies in multiple spectral bands, managers can identify areas of algae growth, pollution, or sedimentation. This information helps in maintaining healthy aquatic ecosystems.

Environmental Impact Studies

Description: Multispectral imaging is invaluable in assessing the ecological impact of construction, mining, or other large-scale projects. It provides baseline data on vegetation and soil health for informed decision-making.

Thermal Missions

Definition and Purpose

Thermal missions use infrared cameras to detect temperature differences on the surface of the Earth or within structures. These sensors capture radiated heat rather than reflected light, providing valuable insights into conditions that are not visible in RGB or multispectral imagery. Thermal missions are crucial for applications such as search and rescue, building inspections, and wildlife monitoring.

Key Features

Temperature Gradient Detection: Measures and visualizes differences in heat signatures.

Nighttime Operations: Thermal sensors can operate effectively in total darkness, aiding round-the-clock inspections or missions.

Non-Destructive Testing: Allows operators to identify structural anomalies or water leaks without invasive methods.

Variable Resolution: Image resolution depends on the thermal sensor's specifications; higher-end sensors offer more precise temperature readings.

Applications: Commonly used in firefighting, search and rescue, building inspections, precision agriculture, and wildlife monitoring.

Uses for Thermal Missions

Building and Roof Inspections
Description: Thermal imaging reveals heat loss, insulation gaps, and water intrusions in roofs and walls. Property managers and homeowners use this data to improve energy efficiency and target repairs.

Search and Rescue
Description: Thermal drones can quickly detect heat signatures of missing persons, especially in challenging environments like dense forests or disaster zones. This significantly reduces search times and increases the likelihood of successful rescues.

Firefighting and Fire Monitoring
Description: Firefighters use thermal drones to locate hotspots, track fire spread, and assess extinguished areas. This improves firefighter safety by providing real-time data on the fire's behavior and intensity.

Solar Panel Inspections
Description: Thermal sensors detect hotspots and faulty cells in solar panels. Identifying malfunctioning panels early helps owners maintain system efficiency and prevent potential fire hazards.

Electrical and Power Line Inspections
Description: Overheating in transformers, power lines, or substations can signal impending failure. Thermal drones detect these anomalies, allowing preventive maintenance and reducing the likelihood of outages.

Pipeline and Leak Detection
Description: Thermal imaging identifies temperature anomalies in pipelines carrying liquids or gases. This early detection helps prevent environmental damage and costly pipeline failures.

Livestock and Wildlife Monitoring
Description: Farmers and conservationists use thermal imaging to track animal movement, detect nighttime poaching activity, and monitor herd health. Injured or sick animals often exhibit distinct heat signatures.

Agricultural Water Stress Analysis
Description: Thermal sensors reveal variations in crop canopy temperature, indicating water stress. Farmers can use this information to optimize irrigation schedules and improve water-use efficiency.

HazMat and Industrial Inspections
Description: Thermal drones help identify hot spots, chemical leaks, or overheating equipment in industrial settings. This data allows for proactive maintenance and improved safety protocols.

Disaster Assessment and Recovery
Description: In post-disaster scenarios (earthquakes, tornadoes, hurricanes), thermal drones can locate survivors trapped under debris and assess the integrity of damaged buildings by detecting heat anomalies or water intrusion.

Integrating Specialized Missions into Your Workflow

When deciding on whether to use LiDAR, Multispectral, or Thermal sensors, consider the specific requirements of your project.

LiDAR is the go-to solution for detailed 3D mapping and measurements requiring high accuracy, especially under canopy or in low-light conditions.

Multispectral imaging is essential when the focus is on vegetation health, soil analysis, and environmental monitoring.

Thermal sensors excel in identifying heat signatures for inspections, search and rescue, and firefighting.

Many drone operators find that a combination of these sensors can provide a more comprehensive dataset. By overlaying thermal data with multispectral indices or LiDAR point clouds, you can gain deeper insights and make more informed decisions. As always, the choice of sensor should align with the mission's objectives and regulatory considerations.

Chapter 11: Mapping Workflow with a Drone

Drone photogrammetry missions are intricate operations that blend cutting-edge technology with meticulous planning and execution. Earlier chapters detailed the hardware and software foundations of drone mapping; now, we turn to the practical orchestration of a survey mission. This chapter provides an in-depth workflow for planning, preparing, executing, and processing data, leveraging Real-Time Kinematic (RTK) and Post-Processed Kinematic (PPK) technologies, individually, or combined, for precision. Spanning initial objectives to final deliverables, this process is adaptable to diverse applications, from topographic surveys to infrastructure monitoring. A detailed section on mission planning with DJI Pilot 2 follows, offering a hands-on guide for a widely used platform. With this workflow, operators can achieve high-quality results efficiently, meeting client needs with confidence.

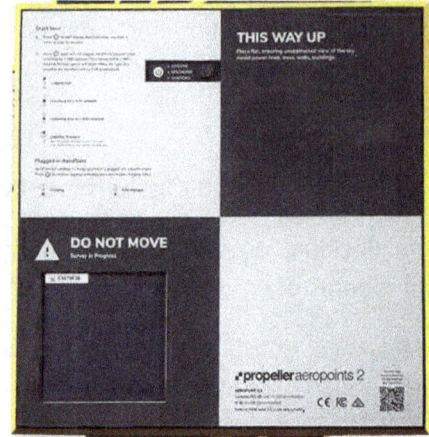

Drone Survey Mission Workflow

Planning

Defining Objectives and Scope

A mission's success hinges on a clear purpose. Start by identifying the survey's goal—topographic mapping for land development, boundary surveys for legal disputes, archaeological documentation, or construction progress tracking. Each objective shapes data requirements: resolution (e.g., 2 cm/pixel for detailed orthomosaics), coverage area (e.g., 50 acres), and deliverables (e.g., 3D models, volumetric calculations). For instance, a mining survey might prioritize stockpile volumes, while solar inspections demand thermal imagery. Document these specifics to guide equipment and flight choices.

Selecting the Survey Area

Using tools like Google Earth or USGS maps, delineate the survey area's boundaries. Assess geographical constraints—rivers, dense forests, steep slopes, or urban zones—that may influence flight paths or altitude. Consider restricted areas (e.g., near airports) and plan buffer zones to avoid legal or safety issues. For a 50-acre rural site, note elevation changes requiring Terrain Follow capabilities to maintain consistent ground sampling distance (GSD).

Regulatory Compliance and Coordination

Compliance is non-negotiable. In the U.S., verify airspace status via the FAA's LAANC system for instant authorizations or request a Part 107 waiver for complex operations. Monitor Temporary Flight Restrictions (TFRs) via apps like AirMap—e.g., avoiding wildfire zones in California. Secure landowner permission for private property and notify local authorities if operating near public spaces. Coordinate with stakeholders (e.g., construction managers) to schedule flights during optimal conditions—midday light, winds below 15 mph—ensuring data quality and safety.

Preparation

Equipment Selection

Selecting the optimal drone and sensor pairing is essential for achieving mission-specific precision in photogrammetry workflows. Real-Time Kinematic (RTK) drones, such as the DJI Matrice 350 RTK or Mavic 3 Enterprise, deliver immediate GPS corrections through integrated GNSS systems, achieving theoretical accuracies of 1 cm plus 1 ppm when connected to a Continuously Operating Reference Station (CORS) network, ideal in areas with reliable internet or local base stations like the Tremble R series or the Emlid RS Series. Base station placement can enhance outcomes: siting one over a known monument can ensure both relative and local accuracy, while a surveyor-established point can ensure absolute accuracy. In environments where RTK connectivity falters due to the absence of CORS coverage or base station limitations (internet is usually required) Post-Processed Kinematic (PPK) techniques offer a robust alternative. Drones like the Inspired Flight IF800 Tomcat, equipped with high-accuracy GNSS receivers, log raw satellite data, and when used with properly placed GCPs for post-flight correction, ensure precision without real-time signals. Hybrid RTK/PPK systems, which drones like the Vision Aerial Switchblade and the DJI M350RTK are capable of, provide versatility by supporting both immediate corrections and post-processed refinement. When paired with sensors like the Zenmuse P1 (45MP, 35mm lens) for photogrammetry or LiDAR units like the Rock Robotic R3Pro or DJI Zenmuse L2, these drones integrate seamlessly with software, aligning hardware performance with the demands of geospatial analysis.

Flight Route Design

Automated flight planning, facilitated by software like DJI Pilot 2 or DroneDeploy, ensures comprehensive coverage and precision in drone mapping missions. Flight paths are defined by setting waypoints to encompass the target area, with altitude adjustments tailored to achieve the desired Ground Sample Distance (GSD)—for instance, flying at 200 feet with a 35mm lens yields a resolution of 0.8 cm per pixel. Optimal image stitching requires a front overlap of 75% and a side overlap of 70%, providing sufficient data for robust photogrammetric processing. On uneven terrain, enabling Terrain Follow—supported by a preloaded Digital Elevation Model (DEM) such as ASTER or real-time RTK sensors— maintains consistent altitude and image quality. Mission design varies by site: a 50-acre plot benefits from a grid pattern for thorough coverage, whereas a linear highway survey aligns with a streamlined single-pass route. These parameters, when properly configured, maximize data accuracy and operational efficiency.

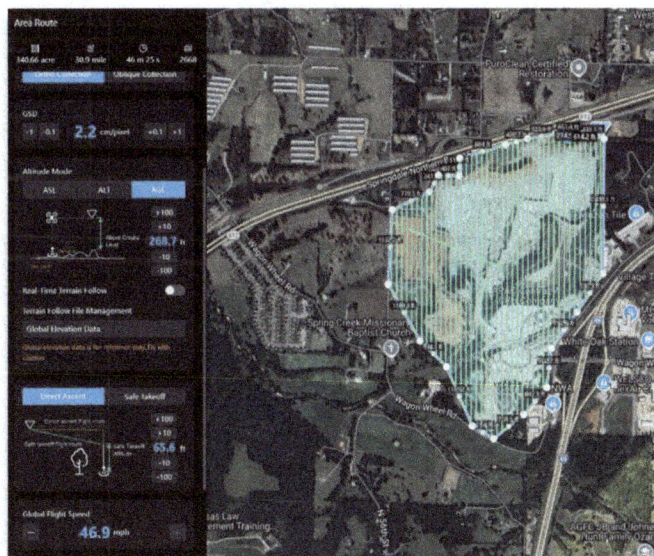

Ground Control Points (GCPs)

Ground Control Points (GCPs) are physical markers strategically placed across a survey site to tie drone captured data to real-world coordinates, significantly enhancing positional accuracy beyond what RTK or PPK alone can achieve. Typically, 5 to 10 high-contrast targets—such as 2x2-foot checkerboard patterns are distributed evenly, with additional points recommended for complex or rugged terrain to ensure robust georeferencing. Several methods facilitate their deployment: a survey-grade GNSS receiver, like the Emlid Reach RX, leverages CORS network corrections, logging each point for 30 to 60 seconds to deliver sub-centimeter precision; alternatively, a local base station, such as the DJI D-RTK 2 positioned over a known monument, pairs with a rover to record GCPs via NTRIP corrections. In scenarios where RTK consistently achieves 3 to 5 cm accuracy, physical GCPs may be omitted, though this risks absolute positional drift over time. Coordinates, typically recorded in the WGS84 datum, can be finalized post-flight, streamlining on-site setup while maintaining geospatial integrity. By anchoring photogrammetric outputs to verifiable ground truth, GCPs elevate the reliability of maps, models, and measurements across diverse applications.

Equipment Checklist

A successful drone mapping mission begins with meticulous equipment preparation. All gear—drones, sensors, base stations, and batteries—must be verified for functionality and fully charged prior to deployment. Battery requirements hinge on mission scope: for instance, the DJI Matrice 350's TB65 battery, with a 41-minute runtime, operates safely at 80% capacity (approximately 33 minutes), covering 40 to 50 acres per flight under typical conditions. To mitigate risks, pack 50% more batteries than calculated along with spares such as propellers and high-capacity SD cards (128 GB minimum) to handle extensive image sets. Firmware and software, including tools like DJI Pilot 2, should be updated in advance to avoid mid-mission failures; nothing disrupts a survey more than an uncharged battery or an untimely software glitch. This proactive approach ensures operational readiness and data integrity across diverse mapping scenarios.

Execution

Pre-Flight Protocols

Thorough pre-flight preparation ensures a seamless mission launch. A detailed checklist guides the process, starting with equipment validation: the drone, sensor, and RTK or PPK systems are tested to confirm connectivity, such as achieving an RTK "FIX" status or verifying stable GNSS logging. Weather conditions are assessed to ensure clear skies, minimal wind, and optimal visibility, all critical for data quality. The team—comprising the pilot, observer, and data manager—is briefed on their roles and the flight plan, aligning efforts for efficiency. The mission begins with a manual launch to evaluate stability and system performance, providing confidence before transitioning to the automated flight path. For a 50-acre site, a single flight may cover the area with sufficient battery life, whereas larger surveys require strategic battery swaps, resuming from the last waypoint to maintain uninterrupted coverage.

In-Flight Monitoring

Continuous oversight during flight is vital to mission success, facilitated by real-time tracking through software like DJI Pilot 2. The pilot monitors telemetry data—altitude, battery levels, and RTK signal strength—to ensure operational stability throughout the survey. RTK missions rely on a consistent connection to a CORS network or a local base station, such as the DJI D-RTK 2, to maintain georeferencing accuracy. In contrast, PPK missions focus on uninterrupted GNSS data logging, preserving raw data for post-flight correction. Should obstacles like trees or structures emerge, the pilot adjusts altitude or speed to safeguard coverage, ensuring image capture aligns with the planned parameters—for instance, maintaining 0.7-second intervals in the Zenmuse P1's Smart Oblique mode to meet GSD and overlap targets. This proactive monitoring guarantees that data collection adheres to the mission's photogrammetric objectives.

Data Processing
Data Management

Effective data handling begins with transferring files to a computer, organized into logical folders by flight segment or type—images, GNSS logs, and telemetry. A 50-acre survey typically generates 500 to 700 images, each tagged with XMP metadata (coordinates, timestamps), demanding structured storage for streamlined access. This foundational step ensures compatibility with subsequent photogrammetric processing.

Processing Steps

Processing varies by GNSS method, yet follows a systematic path to actionable outputs. For RTK missions, images import directly into software like Pix4D or Agisoft Metashape, leveraging real-time corrections to construct an initial sparse point cloud. PPK workflows require an additional step: GNSS logs are refined using tools like Emlid Studio, applying post-processed corrections before integration with imagery. Combined RTK/PPK approaches merge both datasets for maximum precision. From there, a dense point cloud—potentially millions of points—emerges, followed by a 3D mesh and textured model. Final deliverables, including orthomosaics, Digital Elevation Models (DEMs), and Digital Surface Models (DSMs), may be enhanced with GCPs for georeferencing. Quality assurance entails verifying accuracy against GCPs (targeting <5 cm RMSE) and addressing gaps—such as re-aligning images if overlap falters over tall vegetation—ensuring robust, reliable results.

Deliverables
Output Preparation

Final deliverables are tailored to client specifications, enhancing their utility across applications. Export orthomosaics as GeoTIFFs, point clouds as LAS files, and 3D models as OBJs, supplementing with value-added products like contour maps, volumetric reports, or thermal overlays. Compile a comprehensive report—often auto-generated by software—detailing methodology, equipment (e.g., Matrice 350 with Zenmuse P1), processing steps, and accuracy metrics, such as 3 cm horizontal precision. This documentation provides transparency and context, elevating the deliverable's professional impact.

Delivery and Support

Present deliverables in an organized format, whether through a structured folder system or a web platform like DroneDeploy or WebODM. Secure cloud links or physical drives ensure safe transfer, accompanied by user guides explaining data integration—such as importing into ArcGIS for further analysis. Enhance client value with post-delivery support, offering a concise training session (e.g., a 1-hour GIS tutorial) or technical assistance for troubleshooting. This commitment ensures clients fully leverage the data, bridging the gap between raw outputs and actionable insights.

This workflow, adaptable to RTK, PPK, or hybrid approaches, establishes a scalable framework for precision mapping, capable of addressing projects from compact plots to expansive landscapes with consistent reliability.

Mission Planning with DJI Pilot 2

DJI Pilot 2 serves as a robust platform for orchestrating drone missions, powering precise flight planning for models like the Mavic 3 Enterprise (paired with the DJI RC Pro Enterprise controller) and the Matrice 350 RTK (equipped with the DJI RC Plus controller). This section delineates a comprehensive workflow, enriched with practical insights, to ensure missions deliver high-fidelity geospatial data with operational efficiency.

Pre-Mission Preparation

Area Definition and Software Configuration

Mission success begins with defining the survey area within DJI FlightHub 2 (fh.dji.com), leveraging satellite imagery or imported KML files for accuracy. After logging in with a DJI account, create a project and access the "Flight Route Library" to initiate a new route. Select the mission type—"Mapping" for photogrammetric surveys or "Linear Route" for elongated features like highways—then specify the drone and sensor pairing, such as the Matrice 350 with the Zenmuse P1. Assign a descriptive mission name, locate the site via address or coordinates, and designate a Home Point proximate to the intended launch location. While exact precision isn't required at this stage, a close approximation streamlines subsequent parameter adjustments, setting the foundation for a tailored flight plan.

Parameter Configuration

Flight parameters must align with the mission's objectives to optimize data quality. Set the altitude to 200 feet to achieve a Ground Sample Distance (GSD) of 0.8 cm per pixel with a 35mm lens, ensuring fine detail for photogrammetric outputs. Configure overlap at 75% front and 60% side to facilitate reliable image stitching, balancing coverage with processing demands. A flight speed of 8 m/s strikes an effective equilibrium between image clarity and mission duration, while gimbal settings—nadir for 2D orthomosaics or Smart Oblique Capture for 3D modeling—tailor the capture mode to the desired deliverable. Once configured, save the mission and synchronize it to the controller: connect to Wi-Fi, launch Pilot 2, access FlightHub 2 through "Cloud Service," join the project, and download the route from "Cloud" to "Library." Reviewing parameters in editor mode confirms accuracy, as mid-flight adjustments prove far less practical.

Equipment Readiness

Equipment selection and preparation are pivotal to mission reliability. The Mavic 3 Enterprise offers portability for smaller sites, while the Matrice 350 RTK excels in large-scale, precision-driven surveys. Verify the condition of all components: fully charge TB65 batteries (offering 55-minute endurance), test the Zenmuse P1's shutter response, and pack spares—50% more batteries than estimated, along with additional propellers—to mitigate unforeseen interruptions. Firmware updates for both drone and Pilot 2, executed pre-mission, prevent in-field delays, ensuring seamless integration of hardware and software for optimal performance.

Building the Mission

Within Pilot 2's "Flight Route" > "Library," refine the mission by opening the saved plan and enhancing its structure. Define the survey area by drawing boundaries or importing a KML file to establish waypoints, ensuring comprehensive coverage. Activate RTK functionality by connecting to a CORS network—entering NTRIP credentials—or positioning a D-RTK 2 base station over a known monument, verifying a "FIX" status for real-time precision. For complex sites with variable terrain or obstacles, incorporate safety buffers or elevate altitude to maintain data integrity. This meticulous design phase transforms the initial plan into a robust, executable mission tailored to site-specific challenges.

Ground Control Points (GCPs)

Ground Control Points (GCPs) enhance geospatial accuracy by anchoring drone data to real-world coordinates. Distribute 5 to 10 high-visibility markers—positioned at corners, the center, and key elevation shifts—across the site, adapting density to terrain complexity. Post-flight measurement with a survey-grade GPS receiver, such as the Emlid Reach RX via CORS corrections, ensures sub-centimeter precision; alternatively, RTK-equipped drones may suffice for less stringent absolute accuracy needs, though drift risks remain. Record coordinates in the WGS84 datum, ensuring compatibility with downstream photogrammetry software, and integrate them into the workflow for refined outputs.

Flight Execution

Execution demands rigorous pre-flight validation and real-time oversight. Conduct a checklist: confirm favorable weather (clear skies, winds below 15 mph), battery levels exceeding 80%, a stable RTK link, and gimbal calibration for consistent imaging. Launch the drone manually to assess stability, then initiate the automated mission via Pilot 2. Monitor progress through the software's interface and FlightHub 2 on a secondary device (e.g., a laptop), tracking telemetry—battery status, signal strength, and flight path adherence. Upon completion, inspect the drone for wear and transfer data from the SD card, preserving the mission's integrity for processing.

Data Processing and Delivery

Post-flight, download imagery and import it into photogrammetry software alongside GCP coordinates, if utilized. Processing entails aligning images, generating point clouds, constructing a mesh, applying textures, and producing orthomosaics and Digital Elevation Models (DEMs). Export deliverables in GIS-compatible formats—such as GeoTIFF, LAS, or OBJ—and enhance them with a detailed report outlining methodology, equipment specifications (e.g., Matrice 350 with Zenmuse P1), and accuracy metrics.

Client training, such as a concise QGIS demonstration, further elevates the deliverable's utility, ensuring seamless integration into their workflows.

This DJI Pilot 2 workflow complements broader mission frameworks, offering a platform-specific roadmap that balances precision, efficiency, and adaptability. It empowers operators to harness drone technology with confidence, delivering geospatial insights tailored to diverse applications.

Chapter 12: Transforming Drone Data for Professional Use

Drone technology has reshaped data collection across industries such as surveying, construction, environmental monitoring, and urban planning, delivering unprecedented access to aerial perspectives. Yet, the true power of drone-captured LiDAR and photogrammetry data emerges only through its transformation into actionable, survey-grade deliverables. In Chapter 7 and 8, we explored Photogrammetry and LiDAR processing software. In this chapter we will take a brief look at five industry-leading software solutions, Pix4Dmapper, Global Mapper, Bentley ContextCapture, Agisoft Metashape, and Trimble Business Center, that excel in converting raw drone outputs into GIS and CAD-compatible formats. While these platforms represent a robust cross-section of capabilities, they are but a fraction of the hundreds of specialized tools available, each offering unique strengths. This overview aims to illuminate their features and applications, empowering professionals to evaluate these solutions against project-specific needs and explore additional options within the broader market. Far from exhaustive, this chapter provides a focused lens on widely adopted tools, none of which were provided to the author, sponsored this work, or, in most cases, are aware of its publication as of April 2025.

Pix4Dmapper

Overview and Applications

Pix4Dmapper stands as a versatile photogrammetry platform, blending user-friendly design with powerful processing to deliver survey-grade outputs from drone or terrestrial imagery. Its automated workflows streamline the journey from raw data to actionable deliverables, making it accessible to both novices and seasoned professionals. In construction, it generates precise digital surface models (DSMs) and orthomosaics for site analysis; in agriculture, it produces NDVI maps to monitor crop health; and in infrastructure planning, it supports detailed urban and rural mapping. Integration with Ground Control Points (GCPs) ensures centimeter-level accuracy, cementing its utility for high-stakes surveys.

Key Features and Capabilities

- Automated photogrammetry with Structure-from-Motion (SfM) for point cloud generation.
- GCP support for enhanced georeferencing precision.
- Export options including GeoTIFF, LAS, DXF, SHP, and OBJ.
- Batch processing to manage extensive datasets efficiently.
- Mobile app for real-time, on-site data capture and validation.

Advantages and Limitations

Pix4Dmapper's streamlined workflow and high accuracy with GCPs make it a standout, bolstered by flexible licensing (perpetual or subscription). However, its computational demands can strain hardware when processing large datasets, and mastering its advanced features may require significant training.

Global Mapper

Overview and Applications

Global Mapper emerges as a GIS titan, adept at processing a wide array of spatial data, from LiDAR to photogrammetry outputs. Its versatility and extensive file format support make it a go-to for environmental monitoring, urban planning, and forestry. The specialized LiDAR module enables point cloud classification, feature extraction, and volumetric analysis, while its perpetual licensing appeals to organizations seeking long-term value. Professionals managing complex datasets benefit from its efficient rendering and robust analytical tools.

Key Features and Capabilities

- Support for over 300 file formats, including LAS, GeoTIFF, DXF, and SHP.
- LiDAR module for advanced point cloud manipulation and classification.
- Tools for contour generation, slope analysis, and watershed modeling.
- Integrated 3D visualization for immersive data exploration.
- Batch processing for seamless handling of large-scale projects.

Advantages and Limitations

Its ability to manage diverse data types and modular licensing—allowing users to tailor features to needs—sets it apart, though a steeper learning curve and additional costs for advanced modules may challenge new users.

Bentley ContextCapture

Overview and Applications

Bentley ContextCapture specializes in crafting photorealistic 3D reality meshes from aerial and terrestrial data, excelling in large-scale endeavors like urban infrastructure and heritage preservation. Its integration with Bentley's ecosystem (e.g., MicroStation, ProjectWise) fosters collaborative workflows for enterprises, while cloud-based processing alleviates local hardware burdens. Though subscription-focused, its capacity for detailed, visually striking models justifies the investment for complex projects.

Key Features and Capabilities

- High-resolution 3D mesh generation from diverse imagery sources.
- Advanced georeferencing for precise spatial alignment.
- Cloud processing to tackle massive datasets efficiently.
- Seamless integration with Bentley's engineering software suite.

Advantages and Limitations

ContextCapture delivers stunning accuracy and scalability, but its subscription-heavy model and cost may deter smaller firms, with perpetual licenses available only upon request.

Agisoft Metashape

Overview and Applications

Agisoft Metashape offers a photogrammetry solution optimized for detailed 3D modeling, orthomosaic creation, and multispectral analysis, serving archaeology, forestry, and agriculture. Its precision, enhanced by GCP integration, and support for Python scripting for automation make it a flexible choice. The perpetual licensing model provides cost-effective ownership for small to mid-sized teams, balancing capability with affordability.

Key Features and Capabilities

- SfM-based photogrammetry for dense point cloud production.
- GCP integration for survey-grade accuracy.
- Multi-camera and multispectral data processing capabilities.
- Python scripting for customized workflow automation.
- Export formats including GeoTIFF, LAS, OBJ, and SHP.

Advantages and Limitations

Metashape shines with its multispectral support and cost efficiency, though it struggles with extremely large datasets, and advanced customization may demand programming skills.

Trimble Business Center (TBC)

Overview and Applications

Trimble Business Center (TBC) caters to surveyors and GIS professionals requiring uncompromising accuracy and integration. It processes data from drones, GNSS, and terrestrial scanners, excelling in corridor mapping and construction monitoring. Advanced modules for point cloud analysis and terrain modeling, paired with Trimble hardware compatibility, ensure seamless workflows, while licensing options balance subscription updates with perpetual basics.

Key Features and Capabilities

- Point cloud classification, terrain modeling, and feature extraction tools.
- Integration with Trimble hardware and survey ecosystems.
- Export formats including DXF, DWG, GeoTIFF, and SHP.
- Specialized modules for aerial photogrammetry and corridor analysis.
- Comprehensive data adjustment and refinement capabilities.

Advantages and Limitations

TBC's precision and hardware synergy are unmatched, but its complexity requires training, and advanced features often tie to subscriptions.

Licensing Overview for Drone Data Processing Software

The following table encapsulates the licensing models for these platforms, offering a snapshot of their flexibility and cost considerations:

Software	Subscription Model	Perpetual License Availability	Notable Considerations
Pix4Dmapper	Yes	Yes	Perpetual excludes updates unless maintenance is paid.
Global Mapper	No	Yes	Optional maintenance for updates.
Bentley ContextCapture	Yes	Limited	Primarily subscription; perpetual on request.
Agisoft Metashape	No	Yes	Paid upgrades for major releases.
Trimble Business Center	Yes	Limited	Advanced features often tied to subscription.

Conclusion

The software solutions explored in this chapter each play distinct roles in transforming drone data into professional deliverables. From ContextCapture's visually rich meshes to Global Mapper's modular GIS prowess, these tools cater to diverse needs, budgets, and licensing preferences. Professionals must weigh project requirements, be it precision, scale, or integration, against these platforms' strengths to select the optimal fit, recognizing that this overview merely scratches the surface of a vast software landscape. Together, they exemplify the power of drone data to drive actionable data, seamlessly bridging raw inputs to GIS and CAD workflows with precision and purpose.

Chapter 13: Practical Applications and Case Studies

Drone mapping has transformed industries by enabling rapid and precise data collection from the air. However, the real power of drone technology lies not just in capturing data but in turning it into actionable insights. This is achieved through the seamless integration of various software tools that guide the process from mission planning to professional-grade deliverables.

The drone mapping workflow involves several stages, each reliant on specialized software to ensure accuracy and efficiency. Mission planning tools like FlightHub 2, Pix4D Capture and DroneDeploy are used to design precise flight paths, ensuring systematic data collection. Processing tools such as Pix4Dmapper and Agisoft Metashape then transform raw drone data into refined outputs like orthomosaics, 3D models, and digital terrain models. Ancillary software, including Global Mapper and Trimble Business Center, further enhances the workflow by performing advanced data analysis and converting processed outputs into formats compatible with GIS and CAD systems.

The key to success lies in how these software tools integrate with one another, forming a cohesive ecosystem that bridges the gap between raw data collection and actionable deliverables. This integration ensures data compatibility, reduces processing time, and enhances the accuracy of outputs, making it invaluable for industries such as urban planning, construction, and environmental management.

This chapter explores the intricate workflows created by these software tools, demonstrating how they work together in real-world applications. Through practical use cases such as urban planning, construction monitoring, and search-and-rescue operations, the chapter illustrates the transformative potential of drone mapping when powered by integrated software solutions.

1. Agricultural Multispectral Mapping

Practical Application

Agricultural multispectral mapping uses drones equipped with specialized cameras to capture data across multiple wavelengths, including visible light and near-infrared (NIR). This allows farmers to assess crop health, monitor soil conditions, and optimize resource use. Drones fly over fields, collecting high-resolution images that are processed into vegetation indices like the Normalized Difference Vegetation Index (NDVI). These indices highlight areas of stress, nutrient deficiency, or pest infestation, enabling precision agriculture. Compared to traditional methods—walking fields or relying on satellite imagery—drones offer real-time, detailed insights at a fraction of the cost and time. They can cover hundreds of acres in a single flight, providing farmers with actionable data to adjust irrigation, fertilization, or pest control strategies, ultimately boosting yields and reducing waste.

The process involves drones with multispectral sensors, such as the DJI P4 Multispectral, paired with photogrammetry software like Pix4Dfields. Flights are planned to ensure overlap in images (typically 70–80%), which are then stitched into orthomosaic maps. These maps reveal patterns invisible to the naked eye, like early signs of disease or waterlogging. The technology's scalability makes it ideal for small family farms and large agribusinesses alike, while its affordability—versus hiring manned aircraft—democratizes access to advanced farming tools.

Case Study: Optimizing Soybean Yields in Iowa

In 2023, a 500-acre soybean farm in Iowa adopted drone multispectral mapping to address inconsistent yields. The farmer, facing drought conditions, partnered with a local drone service provider using a WingtraOne drone equipped with a MicaSense RedEdge-MX sensor. Over two days, the drone mapped the entire farm, capturing multispectral data at 8 cm/pixel resolution. Processed through Pix4Dfields, the resulting NDVI maps revealed patches of low vigor in the northwest quadrant, correlating with poor irrigation coverage. Soil moisture analysis further indicated compacted areas hindering root growth.

The farmer adjusted irrigation patterns, targeting the affected zones with an additional 20% water volume, and applied a localized nitrogen boost. By harvest, yields in the treated areas increased by 15% compared to the previous year, adding $12,000 in revenue. The drone survey cost $2,500, yielding a clear return on investment. This case demonstrated how multispectral mapping can pinpoint inefficiencies, turning data into profit while conserving resources—a compelling example of precision agriculture in action.

2. Construction Monitoring of Large-Scale Projects
Practical Application

Drones revolutionize construction monitoring for large-scale projects like apartment complexes, shopping malls, or restaurants by providing real-time aerial oversight. Using photogrammetry, drones capture overlapping images of a site, which are processed into 2D orthomosaics and 3D models. These deliverables track progress, verify compliance with blueprints, and identify issues like structural misalignment or material stockpiles. Flights can be scheduled weekly or daily, offering a dynamic view of sites that ground-based surveys can't match. This reduces the need for personnel to navigate hazardous areas, cuts inspection time, and improves communication among stakeholders via shareable digital assets.

Drones like the DJI Matrice 300 RTK, paired with high-resolution cameras (e.g., Zenmuse P1), enable centimeter-level accuracy. Software such as DroneDeploy or Bentley ContextCapture processes the data, allowing project managers to measure distances, calculate volumes (e.g., earth moved), and overlay designs onto models using Building Information Modeling (BIM). This application shines in managing timelines, budgets, and safety on sprawling developments.

Case Study: Shopping Mall Expansion in Texas

In 2024, a construction firm overseeing a 200,000-square-foot shopping mall expansion in Austin, Texas, employed drone photogrammetry to monitor progress. Using a DJI Matrice 300 RTK, the team

conducted bi-weekly flights, capturing 1,500 images per session at 2 cm/pixel resolution. Processed in DroneDeploy, the data produced 3D models and orthomosaics, revealing a 10-meter discrepancy in the foundation layout within three weeks of groundbreaking—an error missed by ground crews.

Correcting this early saved an estimated $50,000 in rework costs and kept the project on its 18-month schedule. The drone also tracked material stockpiles, identifying a 15% over-order of concrete, prompting adjustments that cut waste by $20,000. Total drone service costs were $8,000 over six months, a fraction of traditional surveying expenses. This case underscores how drones enhance precision and efficiency in complex builds, All reducing costs and improving output.

3. Solar Panel Thermal Inspections
Practical Application

Solar panel thermal inspections leverage drones with thermal cameras to detect inefficiencies in photovoltaic arrays. Hotspots, caused by defects, dirt, or shading, reduce energy output and can lead to long-term damage. Drones equipped with sensors like the FLIR Vue Pro fly over solar farms, capturing thermal and RGB images. Photogrammetry software stitches these into maps, pinpointing anomalies for targeted maintenance. This method outperforms manual inspections, which are slow and risky on large or elevated installations, by covering vast areas quickly and safely.

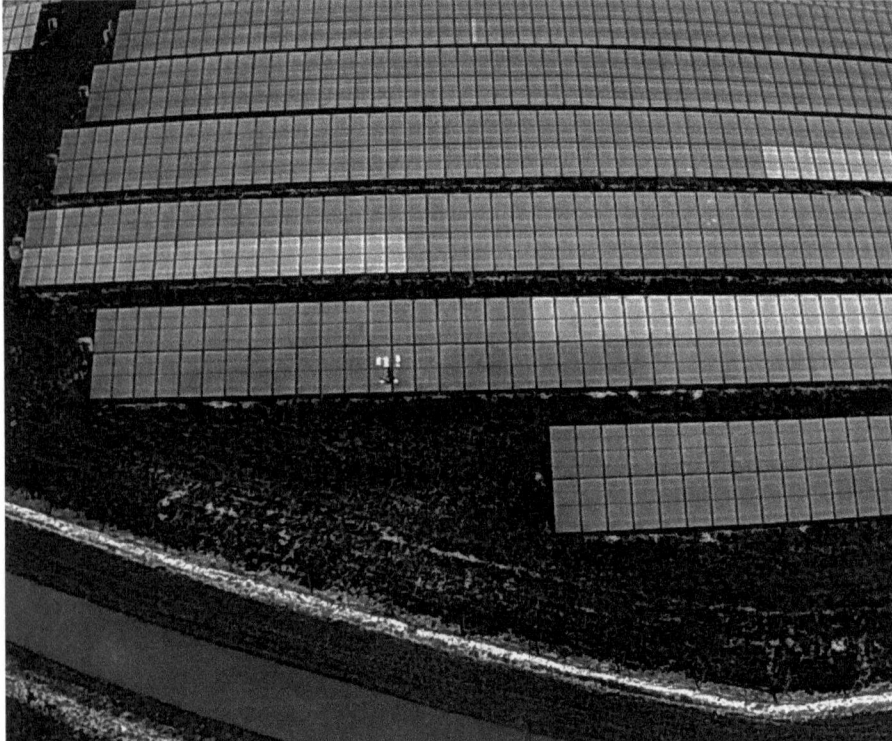

The process requires drones with thermal payloads (e.g., DJI Mavic 3T) and software like Pix4Dinspect to analyze heat signatures. Flights are typically low-altitude (50–100 feet) to ensure resolution, with data revealing issues like faulty cells or wiring failures. This application is critical for renewable energy operators aiming to maximize uptime and efficiency.

Case Study: Solar Farm Maintenance in Arizona

In 2024, a 50-megawatt solar farm in Arizona used drone thermal inspections to address a 5% output drop. A service provider deployed a DJI Mavic 3T, flying over 100 acres in four hours and capturing thermal data at 5 cm/pixel resolution. Processed in Pix4Dinspect, the maps identified 120 panels with hotspots due to dust accumulation and 15 with defective cells. Maintenance crews cleaned the affected panels and replaced the faulty ones within a week.

Post-repair, output rose by 4%, recovering $30,000 in monthly revenue. The drone inspection cost $3,000, versus $15,000 for a manual survey, saving time and labor. This case highlights how thermal drone mapping ensures solar reliability, and is a practical example of increasing production while reducing cost.

4. Large-Scale Topographical and Boundary Surveys
Practical Application

Drones excel in large-scale topographical and boundary surveys, using LiDAR or photogrammetry to map terrain and define property lines. LiDAR-equipped drones (e.g., DJI Matrice 300 with L1 sensor) penetrate vegetation for accurate ground models, while photogrammetry suits open areas with high-resolution cameras. Both produce digital elevation models (DEMs), point clouds, and orthomosaics, delivering survey-grade accuracy (1–5 cm). This replaces labor-intensive ground surveys, reducing time and risk in rugged or expansive landscapes.

Flights are aided by Real-Time Kinematic (RTK) positioning and ground control points (GCPs) for precision. Software like Agisoft Metashape processes the data, enabling land developers, surveyors, and governments to plan infrastructure or resolve disputes efficiently.

Case Study: Boundary Survey in Oregon Wilderness

In 2023, a 1,000-acre forested tract in Oregon required a boundary survey for a land dispute. A surveying firm used a DJI Matrice 300 with an L1 LiDAR sensor, mapping the area in three days versus three weeks for a ground team. The LiDAR data, first processed in Terra, then refined in Agisoft Metashape, produced a DEM with 10 cm accuracy, revealing historical fence lines obscured by trees.

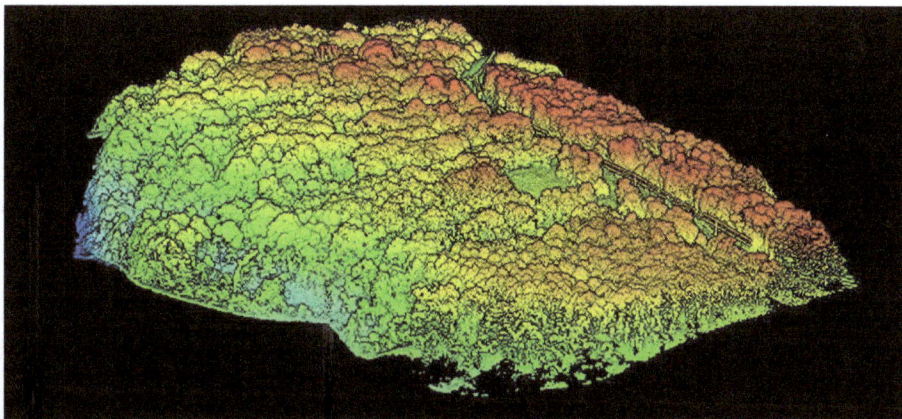

The survey clarified ownership, avoiding a $100,000 legal battle, at a cost $10,000, half the price of traditional methods. This case illustrates drones' power in challenging terrains, a key example of the value of drone use in survey applications.

5. Mapping for Drone Delivery
Practical Application

Drone delivery mapping uses photogrammetry to create detailed 3D models and 2D maps for companies like DroneUp or Zipline. Drones survey urban and rural areas, identifying obstacles (e.g., trees, buildings), optimal flight paths, and landing zones. High-resolution imagery, processed into orthomosaics via tools like Pix4Dmapper, supports route planning and regulatory compliance. This ensures safe, efficient deliveries of goods like medical supplies or retail packages.

Drones like the DJI Phantom 4 RTK provide the precision needed, with flights designed for maximum coverage and overlap. The resulting maps guide autonomous navigation, a growing need as drone delivery scales.

Case Study: Zipline's Medical Delivery in Rwanda

In 2022, Zipline mapped a 50-square-mile region in rural Rwanda to expand medical drone deliveries. Using a DJI Phantom 4 RTK, the team captured 3 cm/pixel imagery over two days, creating a 3D model in Pix4Dmapper. The map identified safe drop zones and avoided power lines, enabling 20 daily flights delivering blood and vaccines.

The mapping cost $5,000, supporting a program that cut delivery times from hours to 15 minutes, saving lives. This case exemplifies drone mapping's role in logistics. A benefit that not only increased productivity while reducing costs, in this case, it saved lives, and continues to do so to this day.

Map generated for a large store in Arkansas currently evaluating delivery operations models (July 2024)

6. Traffic Flow Analysis
Practical Application

Traffic flow analysis uses drones to monitor road networks, capturing aerial video and stills processed into maps or models via photogrammetry. Drones hover over intersections or highways, collecting data on vehicle density, speed, and congestion. Software like DroneDeploy generates orthomosaics or time-lapse analyses, helping planners optimize traffic signals, road designs, or emergency responses. This beats ground sensors or manned aircraft in cost and flexibility.

Drones with stabilized cameras (e.g., DJI Mavic 3E) ensure clear footage, even in dynamic conditions. The application aids urban planning and safety without disrupting traffic.

Case Study: Congestion Study in Los Angeles

In 2024, Los Angeles officials used a flight of drones to analyze a 5-mile stretch of congested freeway. Over three hours, the drone flights recorded peak-hour traffic, and DroneDeploy processed the data into a density map. It revealed a bottleneck at an underused exit, prompting signal adjustments that cut delays by 20%.

The $2,000 survey outperformed a $10,000 ground study, offering real-time insights. This case shows drones' value in urban mobility, increasing efficiency with simple, minor adjustments.

7. Exterior Building and Parking Lot Analysis for Repairs
Practical Application

Drones assess exterior building conditions and parking lots for repairs, using photogrammetry and

thermal imaging to detect damage (e.g., cracks, leaks) or wear (e.g., potholes). High-resolution cameras capture facades, roofs, and pavement, while thermal sensors identify water ingress or heat loss. Processed into 3D models via Pix4Dinspect, the data guides maintenance, reducing scaffolding or manual inspection costs.

Compact drones like the DJI Mavic 3T suit urban settings, offering quick deployment and detailed outputs for property managers.

Case Study: Office Building in Chicago

In 2023, a Chicago property manager used a DJI Mavic 3T to inspect a 20-story office building and its 2-acre parking lot. A two-hour flight produced a 3D model and thermal map, identifying roof leaks and 15 parking lot potholes. Repairs, costing $25,000, were prioritized based on the $1,500 survey, avoiding $5,000 in traditional inspection fees.

This case highlights drones' efficiency in facility management, As large facilities begin to age, the need for continued, cost effective, and detailed inspections will only grow.

8. Volumetric Analysis
Practical Application

Volumetric analysis using drone photogrammetry is a game-changer for industries managing bulk materials, such as rock quarries, sand pits, or gravel operations. Drones equipped with high-resolution cameras or LiDAR sensors fly over sites, capturing detailed imagery or point clouds that are processed into 3D models and digital elevation models (DEMs). These models enable precise calculations of material volumes—whether stockpiles, excavated pits, or remaining reserves—offering a fast, accurate alternative to traditional methods like ground-based laser scanning or manual measurements with tape and theodolites. For quarry operators, this means real-time inventory tracking, better production planning, and improved financial forecasting without the labor-intensive downtime of older techniques.

The process starts with a drone like the DJI Matrice 300 RTK or Phantom 4 RTK, fitted with a 20-megapixel camera or a LiDAR unit like the Zenmuse L1. Flights are planned with tight grids—often 65–75% image overlap—and flown at low altitudes (50–150 feet) to achieve resolutions of 1–3 cm/pixel, critical for capturing the irregular contours of stockpiles or quarry faces. Ground control points (GCPs) or RTK ensure survey-grade accuracy, typically within 2–5 cm. Software such as Pix4Dmapper, Agisoft Metashape, or DroneDeploy then processes the data, generating a 3D mesh or point cloud. By comparing this to a base surface (e.g., the quarry floor), the software calculates cut-and-fill volumes—how much material has been removed or added since the last survey.

This application shines in its efficiency: a 50-acre quarry can be mapped in a few hours, versus days for a ground crew trudging through rough terrain. The resulting data supports inventory management—knowing exactly how many tons of rock are on hand—while also informing extraction rates and equipment allocation. Safety improves, too, as drones eliminate the need for workers to climb unstable piles or navigate active blast zones. For businesses, the cost-benefit is clear: a $1,000 drone survey replaces a $5,000 traditional effort, with results available in a day rather than a week. As drone technology advances, integrating thermal or multispectral sensors could even assess material quality, adding another layer of value.

Case Study: Rock Quarry Inventory in Colorado

In late 2024, a mid-sized rock quarry in Colorado faced inventory discrepancies impacting sales forecasts. The operator, supplying aggregate for road projects, needed accurate stockpile volumes across a 75-acre site with 20 distinct piles of aggregate and limestone. Partnering with a drone service, they deployed a DJI Matrice 300 RTK with a Zenmuse P1 camera, completing a full survey in four hours. The drone captured 2,000 images at 2 cm/pixel resolution, bolstered by five GCPs for precision.

Processed in Pix4Dmapper, the data produced a 3D model and DEM, calculating a total stockpile volume of 150,000 cubic yards—10% less than the operator's rough estimate from ground measurements. Individual pile volumes ranged from 2,000 to 15,000 cubic yards, revealing two piles nearly depleted, prompting a shift in extraction focus to meet a looming contract deadline. The survey also assessed a recent excavation pit, showing 25,000 cubic yards removed over two months, aligning production with demand forecasts. Converted to tons (assuming 1.6 tons per cubic yard for granite), the quarry confirmed 240,000 tons on hand, refining a sales pitch that secured a $1.2 million deal.

The drone survey cost $1,800, versus $7,000 for a week-long traditional survey, and delivered results in 24 hours. Regular monthly flights were later scheduled, building a time-series dataset that tracked inventory trends and optimized blasting schedules, cutting idle time by 15%. This case demonstrates how volumetric analysis via drones turns aerial data into a strategic asset, offering a compelling look at industrial applications with measurable returns.

9. Forestry Management with Multispectral Mapping
Practical Application

Forestry management has embraced drone photogrammetry and multispectral mapping as a powerful tool to monitor replanted areas, track sapling growth rates, and assess loss rates with unprecedented precision. Drones equipped with multispectral cameras—capable of capturing data across visible light, near-infrared (NIR), and sometimes red-edge wavelengths—fly over forested regions, collecting high-resolution imagery that reveals the

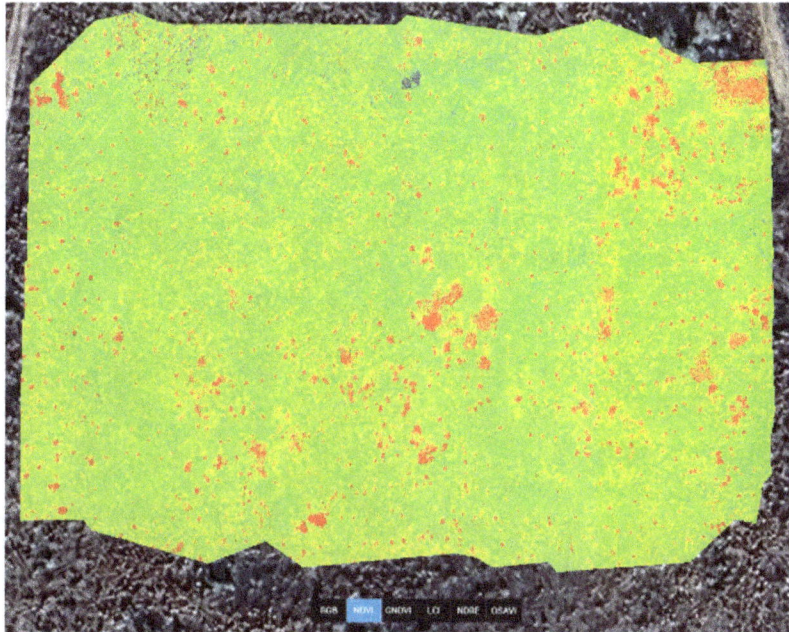

health and density of young trees. This data is processed into orthomosaic maps and vegetation indices, such as the Normalized Difference Vegetation Index (NDVI) or the Enhanced Vegetation Index (EVI), which highlight vigor, stress, or mortality in saplings. For the logging industry, this provides a near-real-time inventory of future timber stocks, enabling better planning for harvests, replanting, and compliance with sustainable forestry regulations.

The process begins with drones like the DJI Mavic 3 Multispectral or the Vision Aerial Switchblade, fitted with sensors such as the MicaSense Alum PT that capture images at resolutions as fine as 1–3 cm/pixel. Flights are typically planned at 150–300 feet above the canopy, with 70–80% image overlap to ensure accurate stitching via photogrammetry software like Pix4Dfields or Agisoft Metashape. These tools generate detailed maps that identify individual saplings, assess their height and canopy development, and flag areas of die-off due to drought, pests, or soil issues. Unlike traditional methods—where biologists conduct labor-intensive 10% sample surveys by trekking through rugged terrain—drones offer a 100% census of replanted zones in hours, not weeks. This comprehensive coverage eliminates the guesswork of extrapolating from small samples, delivering a fuller picture of forest regeneration.

The advantages are manifold. Cost-wise, a drone survey might run $1,000–$3,000 for a 500-acre site, compared to $10,000 or more for a field team's multi-day effort, factoring in salaries, travel, and equipment. Time savings are equally stark: a drone can map an area in a single morning, with data processed by the next day, versus the days or weeks required for manual counts. The multispectral data also detects issues invisible to the naked eye—early pest infestations, nutrient deficiencies, or water stress—allowing forest managers to intervene proactively with targeted replanting, fertilization, or irrigation. For large-scale operations, drones can integrate with Geographic Information Systems (GIS) to track growth trends over years, building a digital archive of inventory as it matures. This scalability and precision make the technology a cornerstone of modern, sustainable forestry, balancing economic goals with ecological stewardship.

Case Study: Reforestation Monitoring in Oregon

In 2024, a logging company in Oregon tasked with replanting a 600-acre clearcut site turned to multispectral drone mapping to monitor sapling survival and growth after a challenging first year. The area, replanted with Douglas fir and western hemlock saplings in 2023, had faced a dry summer and suspected vole damage, raising concerns about inventory losses. Partnering with a drone service provider, the company deployed a DJI P4 Multispectral drone, which flew the site over two days, capturing 3,000 images at 8 cm/pixel resolution across five spectral bands. Processed in Pix4Dfields, the data produced an NDVI map and an orthomosaic, offering a tree-by-tree analysis of the replanted zone.

The results were eye-opening. The drone survey counted 180,000 surviving saplings out of 200,000 planted—a 90% survival rate—far more accurate than the 85% estimate from a biologist's 10% sample the prior month. The NDVI map pinpointed 15 acres in the eastern sector with low vigor, correlating with vole damage confirmed by ground checks, and another 10 acres showing water stress near a poorly drained slope. Armed with this data, the company replanted 20,000 new saplings in the affected zones and adjusted irrigation to prioritize the stressed area, costing $15,000 but preserving long-term inventory value estimated at $150,000 once mature. Growth rates also impressed: healthy saplings averaged 12 inches of height gain, aligning with projections for a harvest in 15–20 years.

The drone survey cost $4,800, a bargain compared to $12,000 for a full biologist team sampling over two weeks, and delivered results in 48 hours. Monthly flights were later scheduled to track progress, building a dataset that refined replanting strategies and satisfied state forestry regulators. This case underscores how multispectral drones revolutionize inventory management, offering your readers a vivid example of how technology outperforms traditional methods in cost, speed, and completeness while supporting sustainable forestry goals.

Conclusion

Drone photogrammetry has emerged as a transformative force across diverse industries, as evidenced by its applications in forestry management, construction monitoring, solar panel inspections, topographic surveys, drone delivery mapping, traffic flow analysis, building maintenance, and volumetric analysis. From the lush replanted forests of Oregon to the bustling highways of California, drones equipped with advanced sensors deliver precision, efficiency, and cost savings that traditional methods struggle to match. The case studies—whether pinpointing sapling survival rates, correcting a shopping mall's foundation, or calculating quarry stockpiles—illustrate how this technology turns raw aerial data into actionable insights, empowering professionals to optimize resources, enhance safety, and drive innovation. Beyond economics, drone photogrammetry supports sustainability, enabling smarter land use and infrastructure management in an era of rapid global change. As the technology evolves, its accessibility and versatility promise to redefine surveying and mapping, making it an indispensable tool for the future. This book has only scratched the surface, but the foundations laid here equip readers to explore and harness drone photogrammetry's vast potential in their own fields.

Chapter 14: Where Are All the Customers?

If you want to get rich in the drone business, it's going to take more than just buying a few drones and the tools that go with them. There are hundreds of thousands of aspiring drone pilots who believe they're going to strike it rich just a few days after getting their Part 107 certification and a Mavic 3 Pro. That's simply not the case. Every year, thousands of drone pilots enter the market, and every year, thousands of pilots leave the market. Building a successful drone business takes more than just a few simple tools and a certification test.

The first thing you need to do is determine what your market is like. What sells in Los Angeles or Seattle may be vastly different from what sells in Tulsa or Wichita. Jobs that once paid $400 to $500 for an hour's work just two years ago are now commonly listed on drone broker websites for $75. Worse yet, these jobs are often more complex and take more time to plan and execute.

Real estate was once a fruitful market for the drone industry, and it still is in a few areas of the country. But in most regions, what was once an hour's worth of work for $250 to $300 has all but evaporated, replaced by photographers who might have gotten their Part 107 just to add the $99 option of aerial video. Ironically, many of these photographers don't even have their licenses. In fact, it's not uncommon to see real estate agents flying these missions themselves, unaware or unconcerned that they need a commercial drone pilot's license simply to save money.

In the United States, the FAA has placed more emphasis on education and less on enforcement when it comes to drone operations. As a result, many individuals knowingly violate the "Personal Recreational Exemption" because they have little fear of punishment. While the FAA will likely begin moving toward stricter enforcement eventually, for the time being, legitimate drone pilots must contend with the reality that many available jobs are being taken by those who cannot legally perform them.

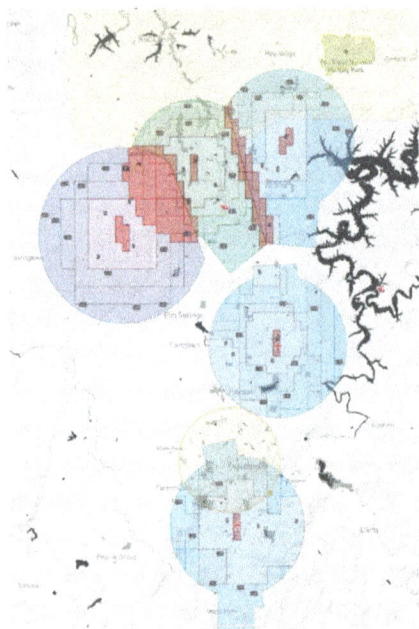

There are also ancillary issues to consider, such as the airspace in which you plan to work. While most major cities have some sort of airspace restrictions, some relatively obscure areas of the country may also have highly restricted airspace, which can limit the profit potential for commercial drone pilots.

Northwest Arkansas is a prime example. While most people may only know it as the headquarters of Walmart, that corner of Arkansas has Class C airspace wrapped in three different Class D airspaces, including one national airport and three other towered airports. Additionally, there are three untowered airports, two of which have active flight schools, all within 20 miles of each other. If you're wondering how this can affect the income of a drone pilot, imagine a 20-acre construction monitoring project that passes through three different grids of restricted drone airspace. Anyone who has waited six weeks for clearance to fly a 20-minute mission will understand the frustration.

But All Is Not Lost

Despite these challenges, there is still significant potential to turn a profit in the drone industry. The key is to find the right niche and adapt to the changing landscape. While certain markets may have become saturated or less profitable, new opportunities are constantly emerging in areas like thermal imaging, construction monitoring, and environmental monitoring. With the right strategy and a focus on specialized services, you can still carve out a profitable space for yourself in this growing industry.

Who Will Benefit from This Chapter?

This chapter is particularly valuable for new drone pilots who are just starting out and are eager to understand how to find and secure clients in a competitive market. It will provide you with a general overview, as well as practical applications, to help you navigate the challenges of establishing yourself in the drone industry.

On the other hand, if you're an experienced drone pilot with a successful business or are already employed by someone else in the industry, you may find that the content of this chapter offers little new information. In that case, you might choose to focus on other chapters that dive deeper into advanced technical services or niche market opportunities.

So, where do you start?
Have a Plan!

Before diving into any specific niche or service offering, it's essential to create a solid business plan. This section will guide you through the key considerations for laying the groundwork for your drone business.

- **Market Research:** Understand the demand in your local area or target market. Research competitors, potential customers, and the types of drone services that are in demand. Consider geographic factors, local industries, and regulatory environments.

- **Regulatory Compliance:** Ensure you are fully compliant with all FAA regulations, including obtaining the necessary certifications, waivers, and insurance. Understand the airspace restrictions in your operating area and how to navigate them.

- **Financial Planning:** Outline your expected income streams, costs (both fixed and variable), and financial goals. Create a budget that includes equipment, marketing, insurance, and other operational expenses.

- **Marketing Strategy:** Develop a marketing plan that includes branding, online presence, networking, and outreach strategies to attract and retain clients.

Focus Areas

This section will explore the various niches within the drone industry, detailing the types of jobs available, the required tools and equipment, customer base, income potential, and legal considerations for each.

1. Thermal Drones

- **Types of Jobs:** Thermal drones are used in a variety of applications due to their ability to detect heat differentials. Typical jobs include building inspections to find insulation leaks or moisture issues, firefighting support to locate hot spots or missing persons in low-visibility conditions, search and rescue operations in challenging terrains, and solar panel inspections to identify faulty or underperforming panels. Additionally, thermal drones are employed in wildlife monitoring and game recovery, where they help in locating animals during nighttime or in dense forests.

- **Tools and Equipment:** Essential tools include thermal imaging cameras, which are critical for capturing temperature variations, and high-resolution RGB cameras for context. Specialized software for thermal analysis is necessary to process and interpret the data, such as FLIR Tools or DroneDeploy with thermal mapping capabilities. Depending on the application, additional equipment might include drones with long battery life, gimbals for stabilized imaging, and night vision capabilities.

- **Customer Base:** The primary customers for thermal drone services include fire departments, search and rescue teams, solar energy companies, construction and building inspection firms, wildlife management agencies, and hunting guides. Each of these sectors relies on the precise and specialized data that thermal drones can provide.

- **Income vs. Cost:** The initial investment for thermal drones and equipment can be significant, often ranging from $5,000 to $20,000 or more, depending on the complexity and capabilities of the equipment. However, the return on investment can be high due to the specialized nature of the services offered. Thermal inspections and emergency services are often billed at a premium, with jobs ranging from $200 to $1,000 per hour, depending on the complexity and urgency of the task. Regular clients, such as solar companies or construction firms, can provide steady, high-paying contracts.

- **Potential Legal Ramifications:** When offering thermal drone services, it's crucial to be aware of privacy and data protection laws, particularly when conducting inspections in residential or commercial areas. Unauthorized surveillance or the capture of thermal data without consent can lead to legal repercussions. Moreover, some industries may require additional certifications or clearances to operate thermal imaging equipment, particularly in sensitive areas like critical infrastructure or public safety operations. Always ensure that your use of thermal drones complies with local, state, and federal regulations, and consider consulting with a legal expert to avoid any potential liabilities.

- **Solar Inspection**

    - **Types of Jobs:** Solar inspections involve using thermal drones to inspect large solar farms or rooftop installations. These drones help identify underperforming panels, detect potential failures, and assess overall system efficiency. The jobs often include routine maintenance checks, post-installation inspections, and pre-purchase evaluations.

    - **Tools and Equipment:** Besides the standard thermal cameras, drones equipped with multispectral sensors are also valuable for detailed analysis of solar panels. Software tools like FLIR Thermal Studio or Raptor Maps are often used to generate reports and analyze data, making it easier to provide actionable insights to clients. High-precision GPS is also necessary to ensure accurate location mapping of faulty panels.

- ○ **Customer Base:** Solar energy companies, maintenance and inspection firms, environmental agencies, and real estate developers who deal with solar installations are the main customers. These clients require regular, detailed inspections to maintain efficiency and comply with regulations.

- ○ **Income vs. Cost:** While the initial cost of equipment is high, the solar industry is growing rapidly, providing a reliable source of income. Inspections are typically charged by the hour or by the panel, with rates ranging from $150 to $500 per hour. The steady need for maintenance and inspection services means that once established, this niche can provide consistent and recurring revenue.

- ○ **Potential Legal Ramifications:** Solar inspections often involve accessing private property or restricted areas, which requires proper authorization and possibly additional insurance coverage. Furthermore, handling sensitive data about the operational efficiency of solar panels can raise concerns about intellectual property and confidentiality. Make sure you have agreements in place that outline data ownership, usage rights, and confidentiality terms. Also, be mindful of any environmental regulations that might apply to the areas you inspect, as solar farms are often subject to strict environmental assessments.

- **Game Recovery**

  - ○ **Types of Jobs:** Game recovery services are highly specialized, often involving the use of thermal drones to locate downed game in dense forests or during low-light conditions. Jobs can also include monitoring wildlife populations, assisting in conservation efforts, or even tracking invasive species.

  - ○ **Tools and Equipment:** Key equipment includes drones with thermal cameras and night vision capabilities, which are essential for spotting animals under challenging conditions. GPS tracking systems are also crucial for marking locations and navigating dense or remote areas. Additionally, rugged drones with long battery life and resistance to environmental factors are preferred in this niche.

  - ○ **Customer Base:** Hunting guides, wildlife management agencies, conservation groups, and private landowners are the primary customers for game recovery services. These clients often require quick and precise results, especially in areas where game retrieval is difficult.

  - ○ **Income vs. Cost:** The market for game recovery is niche, with seasonal demand, but can be quite profitable. Charging rates typically range from $200 to $600 per recovery mission, depending on the difficulty and location. While the initial investment in specialized equipment is moderate to high, the ability to offer a unique service can command premium pricing, especially in regions where hunting is a popular activity.

  - ○ **Potential Legal Ramifications:** Using drones in game recovery involves navigating various wildlife protection and hunting laws. Some regions may restrict or prohibit the use of drones for hunting or game tracking to prevent unfair advantages or disturbances to wildlife. Understanding and complying with these regulations is vital, as violations could result in fines or hunting license suspensions. Additionally, ensure that your operations do not inadvertently interfere with protected species or habitats, which could lead to more severe legal consequences.

2. Mapping

- **Types of Jobs:** Drone mapping services are employed in a wide range of industries, including construction, agriculture, environmental monitoring, and urban planning. Jobs often include creating topographic maps for construction planning, land surveys, flood risk assessments, and infrastructure monitoring. Mapping can also be used for environmental conservation efforts, such as monitoring deforestation or wildlife habitats.

- **Tools and Equipment:** Essential tools include RTK GPS systems for high-precision mapping, drones equipped with high-resolution cameras, and specialized mapping software like Pix4D or DroneDeploy. Photogrammetry tools are also necessary to process images and generate accurate 2D and 3D maps. Depending on the job, additional sensors like LiDAR might be required for detailed terrain mapping.

- **Customer Base:** Key clients for mapping services include construction companies, civil engineers, government agencies, environmental NGOs, and agricultural firms. These clients require precise, reliable data to make informed decisions in their respective fields.

- **Income vs. Cost:** The cost of entry into drone mapping can be significant, particularly if high-end RTK drones and software licenses are required. However, the earning potential is substantial, with jobs ranging from $1,000 to $5,000 per project, depending on the scope and complexity. Mapping services are often in high demand in construction and infrastructure projects, providing long-term contracts and repeat business opportunities.

- **Potential Legal Ramifications:** Many states and countries have strict regulations governing the use of the term "mapping," which is often legally protected and reserved for licensed professionals. Using the term "mapping" in your service offerings without proper licensure could result in legal action. Instead, consider using alternative terms like "orthomosaic generation" or "aerial data collection" to avoid legal complications. Always ensure that your marketing and service descriptions comply with local regulations. Additionally, be cautious about the accuracy and reliability of the maps you produce, as inaccuracies can lead to disputes or legal claims, particularly in construction or legal boundary disputes.

3. Survey

- **Types of Jobs:** Drone surveying services are crucial in construction, mining, and environmental management industries. Common jobs include land boundary surveys, geological surveys, environmental impact assessments, and volumetric analysis for resource management. Surveying can also extend to urban planning and real estate, where accurate land measurements are critical.

- **Tools and Equipment:** Surveying requires precision tools, including survey-grade drones equipped with LiDAR or RTK GPS systems for accurate data collection. High-resolution cameras, photogrammetry software, and specialized surveying tools like total stations may also be necessary, depending on the project's requirements.

- **Customer Base:** Primary customers include civil engineers, construction firms, mining companies, government agencies, and environmental consultants. These clients rely on precise data to ensure the accuracy and safety of their projects.

- **Income vs. Cost:** Entering the drone surveying market involves a substantial investment in high-end equipment and software. However, this investment is offset by the high rates charged for surveying services, which can range from $1,500 to $10,000 per project, depending on the complexity and size of the survey. Long-term contracts with construction firms or government agencies can provide a stable and lucrative income stream.

- **Potential Legal Ramifications:** Surveying is a highly regulated profession in many states and countries, with specific licensing requirements. Unauthorized use of the term "survey" in your business or marketing materials could lead to severe penalties, including fines or legal action. To avoid these issues, it's advisable to use terms like "aerial imaging," "spatial data collection," or "topographic data acquisition." Additionally, partnering with a licensed surveyor or offering your services as part of a broader team that includes licensed professionals can help ensure compliance. Be aware that providing inaccurate survey data, especially if used in legal contexts, could result in liability claims or damage to your professional reputation.

4. Volumetrics

- **Types of Jobs:** Volumetric drone services are used to measure and calculate volumes of stockpiles, earthworks, and other materials, making them essential in industries like mining, construction, and agriculture. Jobs typically include stockpile management, monitoring excavation sites, and ensuring accurate material quantities for billing or resource management.

- **Tools and Equipment:** Essential tools include drones equipped with photogrammetry software for 3D modeling, RTK GPS systems for high-precision measurements, and specialized volumetric analysis software. High-resolution cameras are also necessary to capture detailed images for accurate calculations.

- **Customer Base:** Key clients include mining companies, construction firms, quarries, and agricultural businesses. These customers require precise volumetric data to manage resources efficiently and to comply with regulatory requirements.

- **Income vs. Cost:** The cost of entry into volumetrics is moderate to high, particularly if specialized software and RTK-equipped drones are needed. However, the potential income is substantial, with jobs often billed at $2,000 to $7,500 per project. Volumetric services are frequently required on an ongoing basis, providing opportunities for recurring revenue through long-term contracts with industrial clients.

- **Potential Legal Ramifications:** Volumetric analysis often involves providing critical data used in financial transactions, project planning, and resource management. Inaccuracies in your volumetric calculations can lead to significant financial losses for your clients, potentially resulting in legal claims against your business. It's essential to ensure that your methods and equipment are precise and reliable. Additionally, operating drones in industrial or mining areas often requires special permits or clearances, as well as adherence to strict safety protocols to prevent accidents or unauthorized access to restricted zones.

5. Construction Monitoring

- **Types of Jobs:** Construction monitoring involves using drones to track project progress, conduct site inspections, document milestones, and ensure compliance with safety regulations. Drones are also used to create as-built models, track material usage, and inspect hard-to-reach areas like rooftops or high-rise structures.

- **Tools and Equipment:** Key equipment includes drones with high-resolution cameras for capturing detailed images, photogrammetry software for creating 3D models, and drones with extended flight times to cover large construction sites. Additionally, thermal imaging cameras may be used to inspect buildings for insulation and HVAC issues.

- **Customer Base:** The primary customers for construction monitoring services are construction companies, real estate developers, project managers, and architectural firms. These clients rely on accurate, up-to-date data to make informed decisions and ensure that projects stay on schedule and within budget.

- **Income vs. Cost:** The initial investment in durable drones, high-resolution cameras, and software is significant, but the demand for construction monitoring is steady and growing. Jobs are often billed per flight or per project, with rates ranging from $1,000 to $5,000 depending on the project's scope and complexity. Long-term monitoring contracts can provide consistent income and opportunities for upselling additional services like thermal inspections or 3D modeling.

- **Potential Legal Ramifications:** Providing construction monitoring services places you in a position of significant responsibility, as your data and reports are often used to make critical project decisions. Inaccuracies or delays in providing this information could lead to costly mistakes or project delays, potentially opening you up to liability claims. Moreover, construction sites are heavily regulated environments, and flying drones in these areas requires strict adherence to safety regulations set by organizations like OSHA. Violating these safety standards, either through unauthorized flights or by failing to follow proper protocols, could result in fines or legal action.

6. 3D Orthomosaic Generation

- **Types of Jobs:** 3D orthomosaic generation involves creating detailed 3D models and orthomosaics of terrain, buildings, and other structures for use in planning, analysis, and design. Common jobs include urban planning, construction site modeling, architectural visualization, and environmental monitoring.

- **Tools and Equipment:** Essential tools include high-resolution cameras for capturing detailed images, photogrammetry software for processing and stitching images into orthomosaics, and RTK GPS systems for ensuring accurate georeferencing. Depending on the job, additional sensors like multispectral cameras might be used to add layers of data to the 3D models.

- **Customer Base:** Key clients include architects, urban planners, construction firms, environmental agencies, and real estate developers. These customers require detailed, accurate 3D models for planning, analysis, and presentation purposes.

- **Income vs. Cost:** The investment in equipment and software for 3D orthomosaic generation can be high, but the return on investment is substantial due to the high value of the services offered. Projects are typically billed per model or by square footage, with rates ranging from $2,000 to $10,000, depending on the complexity and scale. The ability to deliver detailed and accurate models can lead to repeat business and long-term contracts with clients in construction and urban development.

- **Potential Legal Ramifications:** When generating 3D orthomosaics, it's critical to ensure the accuracy and integrity of your models, as these are often used in planning, design, and legal contexts. Misrepresentations or inaccuracies in your models could lead to disputes, project delays, or even legal claims if they result in construction errors or zoning violations. It's advisable to clearly communicate the limitations and intended use of your orthomosaics to clients to avoid misunderstandings. Additionally, be mindful of privacy concerns when capturing detailed images of urban areas or private properties, and ensure you have the necessary permissions and adhere to local regulations.

7. Environmental Monitoring

- **Types of Jobs:** Environmental monitoring with drones involves collecting data on deforestation, wildlife habitats, water quality, air pollution, and other environmental factors. Jobs often include monitoring protected areas, conducting biodiversity assessments, tracking pollution sources, and assessing the impact of human activities on ecosystems.

- **Tools and Equipment:** Essential tools include drones equipped with multispectral and thermal sensors for capturing environmental data, environmental monitoring software for data analysis, and drones with long flight endurance to cover large or remote areas. Additional equipment might include LiDAR sensors for mapping forest canopies or terrain features.

- **Customer Base:** The primary customers for environmental monitoring services are environmental NGOs, government agencies, research institutions, and conservation organizations. These clients rely on accurate, up-to-date data to monitor and protect natural resources and biodiversity.

- **Income vs. Cost:** The cost of entry into environmental monitoring can be high, especially if specialized sensors and long-range drones are required. However, the potential for long-term projects and grants provides a stable income stream. Jobs are often billed per project or per data set, with rates ranging from $1,500 to $10,000, depending on the scope and duration. Securing contracts with government agencies or environmental organizations can provide consistent, long-term revenue.

- **Potential Legal Ramifications:** Environmental monitoring often involves working in sensitive or protected areas where strict regulations apply. Unauthorized drone flights or data collection in these areas can lead to significant legal penalties, including fines or confiscation of equipment. It's essential to secure all necessary permits and adhere to environmental protection laws. Additionally, the data collected may be subject to confidentiality agreements or restrictions on its use, particularly if it pertains to endangered species or critical habitats.

8. Inspection Services

- **Roofing Inspections**

  o **Types of Jobs:** Roofing inspections using drones involve inspecting residential and commercial roofs for damage, wear and tear, and potential leaks. Jobs often include post-storm damage assessments, routine maintenance checks, and pre-purchase inspections for real estate transactions.

  o **Tools and Equipment:** Essential tools include drones equipped with high-resolution cameras for capturing detailed images, thermal imaging cameras for detecting moisture or insulation issues, and specialized software for generating inspection reports. Depending on the job, additional tools like zoom cameras might be used for close-up inspections of specific areas.

  o **Customer Base:** The primary customers for roofing inspection services are roofing companies, insurance firms, property managers, and real estate agents. These clients require accurate and timely data to assess the condition of roofs and make informed decisions about repairs, maintenance, or property sales.

  o **Income vs. Cost:** The cost of entry into roofing inspections is relatively low compared to other drone services, making it an accessible market for new drone businesses. Jobs are typically billed per inspection, with rates ranging from $200 to $1,000 depending on the size and complexity of the roof. The high demand for roofing inspections, especially after storms or in regions with harsh weather, provides opportunities for consistent and repeat business.

  o **Potential Legal Ramifications:** When conducting roofing inspections, it's essential to comply with local building codes and safety standards. Unauthorized drone flights over private property can lead to privacy concerns and potential legal action, so always obtain the necessary permissions before conducting an inspection. Additionally, any data collected during the inspection, especially if it reveals significant damage, should be handled with care to avoid potential disputes or liability claims. Make sure to clearly communicate the purpose and limitations of the inspection to clients to manage their expectations.

9. Agriculture and Forestry

- **Forest Management**

  - **Types of Jobs:** Forest management services using drones include monitoring forest health, detecting illegal logging activities, managing fire risks, and conducting biodiversity assessments. Jobs often involve large-scale mapping of forested areas, tracking changes in vegetation, and identifying areas at risk of deforestation or disease.

  - **Tools and Equipment:** Essential tools include drones equipped with multispectral sensors for monitoring vegetation health, thermal cameras for detecting fire risks, and environmental monitoring software for analyzing data. Drones with long flight endurance are necessary to cover large forested areas, and additional tools like LiDAR sensors may be used for detailed terrain mapping.

  - **Customer Base:** Key clients include forestry departments, environmental agencies, logging companies, and conservation organizations. These customers require accurate, up-to-date data to manage forest resources, protect biodiversity, and comply with environmental regulations.

  - **Income vs. Cost:** The cost of entry into forest management services can be high, especially if specialized sensors and long-range drones are required. However, the potential for long-term projects and government contracts provides a stable income stream. Jobs are often billed per project or per data set, with rates ranging from $2,000 to $10,000, depending on the scope and duration. The ability to deliver detailed and accurate data can lead to repeat business and long-term contracts with clients in forestry and conservation.

  - **Potential Legal Ramifications:** Operating drones in forested areas often requires special permits, particularly in protected forests or national parks. It's crucial to comply with environmental protection laws and obtain the necessary permissions before conducting any operations. Additionally, the data collected may be subject to regulations regarding its use, particularly if it pertains to endangered species or sensitive habitats. Violations of these regulations can result in significant fines, legal action, and loss of operating privileges.

- **Wild Game Herd Analysis**

  - **Types of Jobs:** Wild game herd analysis using drones involves monitoring populations of wild game animals, tracking migration patterns, and assessing herd health. Jobs often include aerial surveys of game reserves, population counts, and habitat assessments.

  - **Tools and Equipment:** Essential tools include drones equipped with thermal cameras for detecting animals in dense vegetation, GPS tracking systems for marking locations, and specialized software for analyzing population data. Drones with long-range capabilities are necessary to cover large areas and track migrating herds.

- **Customer Base:** Key clients include wildlife conservation organizations, research institutions, hunting guides, and government wildlife agencies. These customers require accurate, up-to-date data to manage wildlife populations, protect endangered species, and ensure sustainable hunting practices.

- **Income vs. Cost:** The market for wild game herd analysis is niche, with seasonal demand, but can be quite profitable. Charging rates typically range from $1,500 to $5,000 per survey, depending on the difficulty and location. While the initial investment in specialized equipment is moderate to high, the ability to offer a unique service can command premium pricing, especially in regions with significant wildlife populations or where conservation efforts are a priority.

- **Potential Legal Ramifications:** Conducting wild game herd analysis using drones involves navigating a complex web of wildlife protection laws, which vary by region. It's essential to understand and comply with these regulations, particularly those related to endangered species or protected habitats. Unauthorized drone flights in these areas can lead to significant legal penalties, including fines or confiscation of equipment. Additionally, ensure that your operations do not disturb the wildlife, as this can lead to further legal complications and damage your professional reputation.

- **Domestic Herd Monitoring and Management**

  - **Types of Jobs:** Domestic herd monitoring involves using drones to track livestock, monitor grazing patterns, assess herd health, and prevent theft. Jobs often include routine health checks, monitoring of grazing land, and tracking of herds in large ranches.

  - **Tools and Equipment:** Essential tools include drones equipped with thermal imaging cameras for monitoring herd health, GPS tracking systems for locating and managing herds, and multispectral sensors for assessing the quality of grazing land. Drones with long battery life and resistance to environmental factors are preferred for use in large, rugged areas.

  - **Customer Base:** Key clients include ranchers, agricultural cooperatives, livestock associations, and veterinary services. These customers require reliable, real-time data to manage large herds efficiently and ensure the health and safety of their livestock.

  - **Income vs. Cost:** The investment in drones and sensors for domestic herd monitoring is moderate, with the potential for high returns due to the ongoing need for herd management services. Jobs are typically billed per survey or hour, with rates ranging from $200 to $1,000 depending on the size of the herd and the complexity of the monitoring. The demand for such services is steady, providing opportunities for long-term contracts and repeat business.

  - **Potential Legal Ramifications:** Monitoring domestic herds with drones involves operating in areas that are often private property, requiring permissions from landowners. Additionally, it's essential to comply with agricultural regulations and animal welfare laws, particularly if your data is used to make decisions about herd management. Unauthorized flights or mishandling of sensitive data can lead to legal action or loss of client trust, so always ensure you have clear agreements and operate within the bounds of the law.

- **Crop Analysis**

  - **Types of Jobs:** Crop analysis using drones involves monitoring crop health, detecting pests and diseases, assessing irrigation needs, and optimizing fertilization. Jobs often include routine field surveys, analysis of crop stress, and monitoring of crop growth throughout the growing season.

  - **Tools and Equipment:** Essential tools include drones equipped with multispectral sensors for analyzing vegetation health, NDVI cameras for assessing crop vigor, and specialized crop monitoring software for processing and interpreting data. Drones with high-resolution cameras are also used to capture detailed images of crops for further analysis.

  - **Customer Base:** Key clients include farmers, agricultural consultants, agribusiness companies, and research institutions. These customers require accurate, real-time data to make informed decisions about crop management, optimize yields, and reduce input costs.

  - **Income vs. Cost:** The cost of entry into crop analysis services is moderate to high, particularly if specialized sensors and software are required. However, the potential income is substantial, with jobs often billed per acre or data set, with rates ranging from $500 to $2,000 depending on the size of the field and the complexity of the analysis. The growing interest in precision agriculture provides opportunities for recurring revenue through regular crop monitoring services.

  - **Potential Legal Ramifications:** When conducting crop analysis, it's crucial to comply with agricultural regulations, particularly those concerning pesticide use and data privacy. Farmers often rely on the data collected to make critical decisions about crop management, so any inaccuracies or misinterpretations could lead to financial losses and potential legal claims. Additionally, ensure that your operations do not violate privacy laws, especially if the data collection involves neighboring properties.

- **Crop Spraying and Treatment**

  - **Types of Jobs:** Crop spraying and treatment services involve using drones to apply pesticides, herbicides, fertilizers, and other treatments to crops. Jobs often include precision spraying to target specific areas, reducing chemical usage and environmental impact. These services are particularly valuable for large-scale farms looking to optimize their use of chemicals while ensuring uniform coverage and minimizing waste.

  - **Tools and Equipment:** Essential tools include crop-spraying drones equipped with specialized tanks and nozzles for even distribution of chemicals, GPS systems for precision targeting, and software for planning and managing spraying routes. Drones with large payload capacities and long flight times are preferred for covering extensive agricultural areas. It's also crucial to have drones that are resistant to chemicals and designed for agricultural use.

- o **Customer Base:** Key clients include large farms, agricultural cooperatives, agribusiness companies, and government agricultural agencies. These customers require efficient, cost-effective solutions for crop treatment that minimize waste and maximize yields. They often seek services that can scale with the size of their operations and provide consistent, reliable results.

- o **Income vs. Cost:** The investment in crop-spraying drones and related equipment is high, but the potential income is substantial, particularly for large-scale farms. Jobs are typically billed per acre or per treatment, with rates ranging from $50 to $200 per acre depending on the type of treatment and the size of the area. The demand for precision spraying is growing, providing opportunities for consistent, high-volume work during the growing season. The recurring nature of these services also offers opportunities for long-term contracts and stable revenue streams.

- o **Potential Legal Ramifications:** It is important to note that applying pesticides or herbicides using drones typically requires an additional license, often referred to as a Remote Pilot Agricultural Aircraft Operator Certificate or a similar certification depending on the jurisdiction. This is in addition to your Part 107 certification. Regulations vary by state and country, so it's crucial to research and comply with local laws governing the application of chemicals via drones. Failing to obtain the necessary licenses can result in significant fines, legal action, and the suspension of your drone operations. Additionally, there are strict guidelines on how and when pesticides can be applied, especially regarding wind conditions and proximity to sensitive areas like water bodies and residential zones, which must be strictly followed to avoid penalties.

10. Insurance Claims

- **Types of Jobs:** Insurance claim services using drones involve assessing damage after natural disasters, inspecting properties for insurance purposes, and detecting potential fraud. Jobs often include aerial surveys of damaged areas, documentation of property conditions, and assessments of claims related to events like hurricanes, floods, and fires. - **Tools and**

- **Equipment:** Essential tools include drones equipped with high-resolution cameras for capturing detailed images of damage, thermal imaging cameras for detecting issues like water intrusion, and specialized software for generating inspection reports and analyzing data. Additional tools like zoom cameras might be used for close-up inspections of specific areas.

- **Customer Base:** The primary customers for insurance claim services are insurance companies, property owners, legal firms, and disaster recovery agencies. These clients require accurate, timely data to assess claims, determine liability, and process payments efficiently.

- **Income vs. Cost:** The cost of entry into insurance claim services is moderate, with the potential for high returns, especially after major natural disasters. Jobs are typically billed per claim or per inspection, with rates ranging from $300 to $2,000 depending on the size and complexity of the property. The high demand for inspection services after disasters, combined with the ability to offer rapid and accurate assessments, provides opportunities for significant and recurring revenue.

- **Potential Legal Ramifications:** When conducting insurance claim inspections, it's essential to comply with local regulations regarding data collection and privacy. Unauthorized drone flights over private property can lead to legal action, so obtaining the necessary permissions and clearances is critical. Additionally, the data collected during these inspections may be used in legal proceedings, so accuracy and thoroughness are paramount to avoid disputes or liability claims. Ensure that your contracts with insurance companies clearly define the scope of work and your responsibilities to protect against potential legal challenges.

The Data Brokers

Navigating the drone industry can be challenging, especially when it comes to finding consistent work. Data brokers and drone job services companies can play a pivotal role in helping you secure clients and maintain a steady flow of projects. However, understanding how to use these services effectively, along with their advantages and disadvantages, is crucial for your success.

1. Pros of Using Drone Job Services Companies

- **Access to a Broader Market:** Drone job services companies offer a gateway to a wider network of potential clients that you might not be able to reach on your own. These platforms aggregate job postings from various industries, giving you the opportunity to bid on or accept jobs that align with your expertise. This can be particularly beneficial for new drone pilots who are still building their client base.

- **Reduced Marketing Costs:** One of the significant advantages of using these services is the reduction in marketing and client acquisition costs. The platforms typically handle advertising, client outreach, and the initial stages of negotiation, allowing you to focus on the technical aspects of your work. This can save you time and resources, especially if you're not skilled in marketing or prefer to concentrate on flying.

- **Steady Stream of Jobs:** By registering with multiple drone job services companies, you can ensure a more consistent stream of job opportunities. While not every job may be high-paying, the volume of work available can help keep your business running smoothly. These platforms often offer jobs across a variety of industries, allowing you to diversify your income sources and gain experience in different sectors.

2. Cons of Using Drone Job Services Companies

- **Lower Profit Margins:** One of the primary downsides of using drone job services companies is that they typically take a percentage of your earnings as a fee. This can significantly reduce your profit margins, especially for lower-paying jobs. It's essential to factor in these fees when bidding on jobs or setting your rates to ensure that the work remains profitable.

- **Increased Competition:** These platforms often have a large number of drone pilots registered, which means you'll be competing with others for the same jobs. This competition can drive prices down and make it harder to secure higher-paying projects. Additionally, some platforms may favor pilots with more experience or better ratings, which can be a challenge for newcomers trying to break into the market.

- **Lack of Control Over Job Terms:** When working through a job services company, you may have less control over the terms and conditions of the job. Clients using these platforms often have pre-set requirements and budgets, leaving little room for negotiation. This can be limiting if you prefer to tailor your services or set your own prices based on the specific needs of a job.

- **If they don't like the end result, they don't pay**. Perhaps the worst issue when dealing with these platforms is that even if a pilot follows the mission parameters to the letter, occasionally, a dataset will be rejected. In at least one instance, a local pilot flew a mission provided to him through a KMZ file. In many cases, as with this one, a KMZ file will be created by the hiring agency or the client who contracts them. This file can be imported directly into the mission software and will create the map needed. Once the pilot flew the mission and provided the data, it was rejected for flying the wrong path, the very path laid out in the file provided to him. So, beware of many of these agencies as they may make you fix their mistakes.

  - **On a personal note**, of all the agencies the author has worked with, the best by far has been FlyGuys, based out of Louisiana. The pay is generally very fair, and the jobs are normally worth doing. You will communicate directly with the mission coordinator, and in many cases, you will have a direct cell number that will actually get answered.

3. When to Know the Difference

- **Starting Out vs. Established Business:** If you're just starting out in the drone industry, using job services companies can be a valuable way to build your portfolio, gain experience, and establish a client base. However, as your business grows and you develop a reputation, you may find it more beneficial to seek out direct clients who can offer better rates and long-term contracts. Knowing when to transition from relying on these platforms to securing your own clients is key to maximizing your income.

- **Supplementing Workload:** Even established drone pilots can benefit from using job services companies during slower periods. These platforms can help fill gaps in your schedule when direct client work is scarce. However, it's important to be selective about the jobs you accept to ensure they align with your expertise and financial goals.

- **Balancing Risk and Reward:** Consider the types of jobs available on these platforms and weigh the risks versus rewards. For instance, if a job offers a lower payout but requires significant travel or specialized equipment, it may not be worth your time. On the other hand, a steady stream of smaller jobs can help maintain cash flow and keep your business operational during lean times.

4. Drone Pilot Networks

- **Building a Professional Network:** Joining drone pilot networks, both online and offline, can be an invaluable resource for finding jobs, learning about industry trends, and sharing knowledge with peers. These networks often provide access to job postings, industry events, and forums where you can ask questions and get advice from more experienced pilots. Networking can also lead to referrals, partnerships, and collaborations that expand your business opportunities.

- **Reputation and Referrals:** A strong reputation within a drone pilot network can lead to word-of-mouth referrals, which are often more valuable and trustworthy than jobs acquired through anonymous platforms. By actively participating in these networks, offering advice, and sharing your successes, you can position yourself as an expert in your field and attract higher-quality clients.

- **Access to Exclusive Opportunities:** Some drone pilot networks offer access to exclusive job opportunities that aren't available on public job boards. These may include high-profile projects, long-term contracts, or jobs requiring specialized skills. Being part of a network can give you a competitive edge and help you tap into markets that are otherwise difficult to access.

  - **The Drone Pilot Networks**

  - Navigating the drone industry effectively involves connecting with platforms that offer job opportunities tailored to your expertise and availability. These platforms, often referred to as drone pilot networks, differ in how they connect pilots with clients, the types of jobs available, and the methods used for assigning or bidding on work. Here's an overview of several prominent platforms:

  - **1. Zeitview (formerly DroneBase)**

  - **Overview:** Zeitview is a well-established platform that offers a broad range of drone services, from aerial photography and videography to specialized tasks like thermal inspections and mapping. Zeitview connects pilots with clients across various industries, including real estate, insurance, construction, and energy.

  - **Job Assignment Method:** Zeitview typically posts jobs to groups of pilots based on their qualifications, location, and experience for specific missions. These jobs may go quickly if they pay well, or they may sit for long periods. The more pilots using the platform in a given area, the less the pay seems to be. Pilots will receive e-mails notifying them of available flights. From there, it is first come, first serve.

  - **Key Note:** Zeitview recently changed owners along with their name. In the process, much of the work of mission coordination has been moved to overseas companies. This has caused some problems as the personnel are not drone pilots and have little to no knowledge of how drone operations work. In one particular instance, they recently rejected a dataset that was captured with a Matrice M350 using a Zenmuse P1 camera because they believed the P1 was a thermal camera. The mission requirements called for a minimum of a Mavic 3e. However, in the acceptable equipment list, the P1 was listed as an acceptable camera. The trip to fly the mission was 3 hours each way, the staff didn't seem to care that they were making a huge mistake and were about to cost the pilot all of the profit from the mission. In the end, the pilot's solution was to change the metadata on the images to reflect that the mission was performed with a DJI Mavic 3 Enterprise. The employees then accepted the dataset without question, and the issue was resolved.

- **2. FlyGuys**

- **Overview:** FlyGuys is a nationwide drone services company that connects clients with a network of professional drone pilots. The platform focuses on providing a wide range of services, including aerial photography, mapping, inspections, and 3D modeling, to clients in industries like construction, agriculture, and real estate.

- **Job Assignment Method:** Jobs on FlyGuys are typically managed by mission coordinators who assign tasks to pilots based on their qualifications and proximity to the job site. This method ensures that jobs are matched with the most suitable pilots.

- **Key Feature:** FlyGuys emphasizes providing end-to-end support, from project coordination to data delivery, which can be beneficial for pilots who prefer a more structured approach to freelancing.

- **3. Droners.io**

- **Overview:** Droners.io is a popular marketplace where clients can post jobs for drone pilots, covering a wide range of services such as real estate photography, aerial surveys, and special events. The platform is known for its user-friendly interface and extensive job listings.

- **Job Assignment Method:** On Droners.io, jobs are typically posted on a public board where pilots can browse and bid on the projects that interest them. Clients then select the pilot they believe is the best fit based on experience, reviews, and bid price.

- **Key Feature:** The open bidding system allows pilots to choose the jobs that best match their skills and negotiate prices directly with clients, making it a flexible option for freelancers.

- **4. DroneWork**

- **Overview:** DroneWork is a platform that connects clients with professional drone pilots for various missions, including property listings, construction monitoring, and inspections. The platform offers a comprehensive solution, allowing clients to manage the entire process, from hiring to payment, through their system.

- **Job Assignment Method:** DroneWork typically involves job postings where pilots can apply for missions. However, the platform also offers a selection process where clients or coordinators match jobs with pilots based on specific criteria.

- **Key Feature:** The dual approach of open applications and coordinator-based selection gives pilots the flexibility to find work that suits their preferences while ensuring that clients receive qualified professionals.

- 5. Dronegenuity

- **Overview:** Dronegenuity specializes in aerial imaging services and offers drone pilots opportunities to capture high-quality photos and videos for clients in various industries, including real estate, construction, and marketing.

- **Job Assignment Method:** Dronegenuity typically selects pilots for jobs based on their location and experience, rather than using an open job board. This ensures that clients receive reliable and vetted professionals for their specific needs.

- **Key Feature:** The platform's focus on high-quality imaging work makes it an ideal choice for pilots who excel in photography and videography, especially those looking to work on projects that require a polished, professional finish.

- 6. DroneDeed

- **Overview:** DroneDeed is a marketplace that connects freelance drone pilots with clients needing various aerial services. The platform is designed for a vetted community, ensuring that only licensed pilots can join and offer their services.

- **Job Assignment Method:** Similar to Droners.io, DroneDeed allows pilots to browse and apply for jobs posted by clients. Additionally, the platform uses AI-driven recommendations to suggest jobs that match a pilot's experience and skills.

- **Key Feature:** DroneDeed's emphasis on AI-driven job matching and project management tools helps pilots streamline their freelance work, making it easier to manage multiple projects and maintain high standards of service.

5. Local Jobs

- **Tapping into Local Markets:** Focusing on local jobs can be a highly effective strategy, especially if you live in an area with a strong demand for drone services. Local clients often prefer working with nearby pilots for convenience and cost-effectiveness. By building relationships with local businesses, real estate agents, construction firms, and government agencies, you can establish yourself as the go-to drone expert in your area.

- **Leveraging Local Knowledge:** Your understanding of local geography, regulations, and market needs can be a significant advantage. Local clients may require specific knowledge of the area, such as understanding airspace restrictions or being familiar with local landmarks, which can give you an edge over out-of-town competitors.

- **Building Long-Term Relationships:** Working locally allows you to develop long-term relationships with clients who may require regular drone services. Whether it's for real estate listings, construction monitoring, or agricultural assessments, these ongoing relationships can provide a steady stream of income and reduce the need to constantly seek out new clients.

Chapter 15: Where We Are Headed

The Future of Drones in Mapping, Surveying, and Beyond

The drone industry is on the cusp of a transformative era, driven by rapid technological advancements, evolving regulatory landscapes, and expanding use cases. Drones, once primarily tools for aerial photography, are now poised to revolutionize photogrammetry, mapping, surveying, and emerging fields like transportation and delivery. This chapter explores the future directions of drone technology, focusing on its potential in photogrammetry and related applications, while addressing the regulatory, ethical, and environmental challenges that will shape its trajectory. By navigating these opportunities and obstacles, the industry can unlock a smarter, more connected world.

Emerging Technologies and Their Potential Impact

1. Artificial Intelligence and Machine Learning Integration

Artificial Intelligence (AI) and machine learning (ML) are set to redefine drone operations, particularly in data analysis and decision-making. These technologies are already automating the processing of drone-captured data, offering unprecedented efficiency for photogrammetry and mapping professionals.

- **Automation of Data Analysis**: AI algorithms can process vast datasets from drone imagery—orthomosaics, point clouds, and 3D models—identifying patterns, classifying objects (e.g., vegetation, buildings, roads), and detecting anomalies (e.g., structural cracks in infrastructure). For instance, in photogrammetry, ML can automatically segment images to distinguish between natural and man-made features, reducing manual processing time from hours to minutes. Software like Pix4D and Agisoft Metashape are integrating these tools, enabling operators to generate actionable insights faster for applications like topographic mapping or disaster assessment.

- **Real-Time Decision Making**: AI-powered drones equipped with onboard processing can analyze data mid-flight, enabling real-time applications beyond mapping. In search-and-rescue missions, drones can identify human figures or heat signatures in thermal imagery instantly, guiding responders to critical locations. For photogrammetry, real-time analysis could optimize flight paths on-the-fly, adjusting overlap or altitude to capture higher-quality data over challenging terrain.

- **Predictive Analytics**: ML models trained on historical drone data—such as seasonal crop health from multispectral surveys or urban growth patterns—can forecast future trends with high accuracy. In agriculture, predictive analytics could advise farmers on irrigation needs or pest control based on drone-derived vegetation indices. In urban planning, these models could predict infrastructure wear, informing maintenance schedules and reducing costs. This capability extends to photogrammetry, where predictive models could anticipate data gaps, prompting additional flights before processing begins.

2. Advancements in Drone Hardware

Hardware innovations are expanding the capabilities and reach of drones, making them more versatile for mapping and beyond.

- **Next-Generation Sensors**: Compact, high-resolution LiDAR, multispectral, and thermal sensors are revolutionizing data collection. LiDAR drones, like those equipped with the Velodyne Puck or Ouster OS1, can generate detailed 3D point clouds for precise topographic surveys, even in dense vegetation. Multispectral cameras, such as the MicaSense RedEdge-P, enable advanced agricultural mapping by capturing data across multiple spectral bands, while thermal imaging supports infrastructure inspections and wildfire monitoring. These sensors are becoming smaller, lighter, and more affordable, broadening their adoption in photogrammetry.

- **Longer Flight Times**: Battery technology is advancing, with solid-state batteries and hydrogen fuel cells extending flight times from 30 minutes to over an hour. Hybrid drones, combining electric and gas propulsion, can operate for several hours, covering hundreds of acres in a single mission. For large-scale photogrammetry projects—e.g., mapping a 1,000-acre forest—longer flights reduce the need for multiple battery swaps, saving time and resources. Companies like Joby Aviation and Volocopter are pioneering these technologies, initially for transportation but with applications in mapping.

- **Swarm Drones**: Coordinated fleets of drones, or "swarms," use AI to operate collaboratively, covering vast areas efficiently. In photogrammetry, swarm drones could simultaneously map multiple sections of a site, stitching data together seamlessly for large-scale projects like national park surveys or disaster recovery. Swarm technology also supports dynamic re-tasking—if one drone detects an anomaly (e.g., a landslide), others can adjust to focus on that area. This approach is being tested by companies like Skydio and Intel, with potential for integration into mapping workflows by 2030.

3. Cloud and Edge Computing

Computing advancements are enhancing drone data management, enabling faster processing and new applications.

- **Cloud-Based Data Processing**: Cloud platforms like DroneDeploy, Pix4Dcloud, and Microsoft Azure reduce the computational burden on local devices, allowing teams to process terabytes of drone data quickly. For photogrammetry, cloud-based processing can generate orthomosaics and 3D models overnight, enabling near-real-time deliverables for clients in construction or environmental monitoring. Collaboration features also allow multiple stakeholders—engineers, planners, and clients—to access and annotate data remotely, improving project efficiency.

- **Edge Computing**: Onboard processing capabilities, powered by edge computing, allow drones to analyze data mid-flight. In photogrammetry, edge computing could detect data gaps (e.g., poor overlap over a forest canopy) and adjust flight paths instantly, ensuring complete coverage. Beyond mapping, edge computing enables real-time obstacle avoidance in urban environments, critical for delivery drones navigating buildings. Companies like NVIDIA are developing edge AI chips for drones, promising widespread adoption by 2025–2030.

4. 3D and Immersive Mapping

Drones are driving the creation of immersive, interactive mapping solutions, transforming how we interact with geospatial data.

- **Integration with Virtual and Augmented Reality (VR/AR)**: Drone-generated 3D models and orthomosaics can be integrated into VR/AR environments, allowing users to explore sites virtually. For cultural heritage documentation, archaeologists could "walk" through ancient ruins like the Anasazi sites, examining details inaccessible in person. In construction, VR/AR enables architects to visualize building plans overlaid on real-world drone maps, improving design accuracy. Tools like Unity and Unreal Engine are already supporting these integrations, with photogrammetry data as the backbone.

- **Digital Twins**: Drone mapping is central to creating digital twins—virtual replicas of physical assets or environments. In manufacturing, digital twins of factories, built from drone LiDAR data, enable real-time monitoring and simulations. For urban planning, cities like Singapore are using drone mapping to create digital twins for traffic optimization and disaster preparedness. Photogrammetry will play a key role, providing high-fidelity data for these models, with applications expanding to agriculture (e.g., farm twins for yield prediction) by 2035.

5. Autonomous Drone Operations

Autonomy is unlocking new frontiers for drones, extending their reach and utility.

- **Beyond Visual Line of Sight (BVLOS)**: BVLOS operations, enabled by advanced sensors and AI navigation, allow drones to map remote or hazardous areas—mountains, deserts, or post-disaster zones—without human oversight. In photogrammetry, BVLOS drones could survey entire national parks or offshore oil rigs, reducing costs and risks. Regulatory approval for BVLOS is growing, with the FAA and EASA testing frameworks, potentially fully implemented by 2027–2030.

- **Drone Delivery Systems**: While focused on logistics, delivery drone advancements—autonomous navigation, payload capacity, and safety systems—are directly applicable to mapping. Amazon's Prime Air and Wing by Alphabet are developing drones that can navigate complex urban airspace, technologies that could enhance photogrammetry drones' ability to operate in crowded environments. Delivery drones could also carry mapping sensors, expanding their utility for urban surveys or disaster response.

Regulatory and Ethical Considerations
1. Regulatory Challenges

The rapid growth of drone use poses significant regulatory hurdles, requiring innovative solutions.

- **Airspace Management**: As drone numbers increase, advanced traffic management systems (UTM, or Unmanned Traffic Management) are essential to prevent collisions. The FAA's UTM initiative, in partnership with NASA, is developing real-time drone tracking and deconfliction for commercial operations, critical for photogrammetry in urban areas or near airports.

- **Standardization of Rules**: Global harmonization of drone regulations remains elusive. The U.S., EU, and China have distinct requirements for licensing, altitude limits, and BVLOS operations. For photogrammetry, operators must navigate these differences, potentially requiring region-specific training or equipment. International bodies like ICAO are working toward standardization, but progress may take a decade.

- **Certification and Licensing**: As drones gain autonomy and complexity, operators may need advanced certifications—e.g., for BVLOS or swarm operations. Photogrammetry professionals could require specialized training in AI-driven data analysis or sensor calibration, ensuring safety and accuracy in diverse environments.

2. Privacy and Data Security

Drones' data collection capabilities raise significant privacy and security concerns.

- **Data Collection Ethics**: Drones capture high-resolution imagery and sensor data, potentially intruding on private property or recording sensitive information. In photogrammetry, operators must anonymize data (e.g., blurring faces or license plates) and obtain consent for surveys over private land, adhering to GDPR or CCPA guidelines.

- **Data Ownership**: Disputes over data ownership—between operators, clients, and software providers—could hinder collaboration. For photogrammetry projects, contracts must clarify ownership, usage rights, and storage responsibilities, especially with cloud-based processing.

- **Cybersecurity Risks**: Drones' reliance on wireless communication and cloud platforms makes them vulnerable to hacking. Photogrammetry data, often stored in the cloud, could be targeted for industrial espionage or ransomware. Encryption, secure protocols (e.g., AES-256), and regular updates are critical to mitigate these risks.

3. Environmental Impact

Drones' proliferation must be balanced with environmental stewardship.

- **Noise Pollution**: Drone noise, though quieter than manned aircraft, can disrupt urban residents or wildlife. In photogrammetry, quieter drones (e.g., electric models) and noise-reducing propellers are being developed, but regulations may limit operations near sensitive areas like parks or reserves.

- **Wildlife Disturbance**: Mapping in natural habitats—forests, wetlands, or coastal areas—requires minimizing drone impact on fauna. Low-altitude flights or frequent passes could disturb nesting birds or mammals, necessitating guidelines for flight times (e.g., avoiding dawn/dusk) and distances (e.g., 50 meters from wildlife).

4. Ethical Use

Ensuring drones serve society responsibly is paramount.

- **Surveillance Concerns**: Drones' ability to collect data covertly raises ethical questions, particularly in photogrammetry for security or law enforcement. Transparent policies and public engagement can build trust, balancing safety needs with privacy rights.

- **Bias in AI Systems**: As AI drives drone mapping, biases in training data (e.g., over-representing urban areas) could skew results, affecting land use or conservation decisions. Developers must audit algorithms, ensuring fairness and accuracy across diverse environments.

Broader Applications: Transportation, Delivery, and Beyond

Beyond mapping, drones are transforming transportation and logistics, with implications for photogrammetry.

- **Urban Air Mobility (UAM)**: Companies like Joby Aviation and Volocopter are developing electric vertical takeoff and landing (eVTOL) drones for passenger transport, sharing technologies like autonomous navigation and long-range sensors with mapping drones. Photogrammetry could support UAM by mapping urban airspace for safe flight corridors.

- **Last-Mile Delivery**: Amazon, UPS, and DHL are deploying delivery drones, leveraging BVLOS and swarm technologies. These drones could carry lightweight mapping sensors, enabling simultaneous package delivery and urban surveys, reducing operational costs for photogrammetry firms.

- **Disaster Response**: Drones for medical supply delivery (e.g., Zipline) use similar autonomy and sensor tech as mapping drones, enabling rapid deployment for post-disaster surveys—e.g., mapping flood zones or earthquake damage with real-time data.

Conclusion

The future of drone mapping, surveying, and related applications is bright, propelled by technological breakthroughs and expanding use cases. AI, advanced hardware, cloud computing, and autonomy promise to enhance photogrammetry's efficiency and scope, while transportation and delivery innovations offer synergistic opportunities. However, realizing this potential requires navigating regulatory complexities, ethical dilemmas, and environmental concerns. By fostering innovation responsibly—through standardized regulations, robust privacy protections, and sustainable practices—the drone industry can transform how we map, monitor, and move in our world. By 2035, drones may become ubiquitous tools for photogrammetry, urban planning, agriculture, and logistics, building a smarter, more connected future for all.

Acknowledgments

This book emerges as a humble endeavor to stem the tide of misinformation that has long inundated the engineering and drone services industry—a clarion call to elevate discourse and practice in a field too often muddied by half-truths and hasty conclusions. I am not alone in this conviction; there are kindred spirits who, like me, have felt the pressing need for such a work to take flight. To name each soul and every enterprise that has lent its wisdom or encouragement would fill pages beyond counting, yet most of you, I trust, recognize yourselves in these words.

To those who have walked beside me along this journey—whether by offering counsel on these pages, imparting the intricate arts of drone technology over the years, or aiding in the practical mastery of its myriad applications—my gratitude runs deep. Your insights, shared generously across countless conversations and collaborations, form the bedrock of this effort. Know that your contributions ripple outward, promising to fortify our community and, with hope, to quell the persistent swell of errant information. Together, we are building something enduring—a legacy of clarity, competence, and connection.

About the Author

Thomas Dowell is an accomplished Private Pilot as well as a commercial drone pilot with over seven years of experience in aerial mapping and commercial drone operations. Since earning his commercial drone license in 2018, he has collaborated with inspection companies, Real Estate Developers, civil engineers, and surveyors on projects ranging from LiDAR and photogrammetry to volumetrics and RGB and thermal inspections. His expertise spans diverse applications, including multispectral forestry analysis, infrastructure inspections, search and rescue missions, big game recovery, and pet searches.

A private pilot and the son of a pilot, Tom has been captivated by flight since the age of five, when he first took to the skies while sitting in the right seat of his father's Piper Cherokee 180. This lifelong passion, combined with his engineering background, fuels his authoritative yet practical approach to drone technology. Through this book, he aims to share his knowledge with enthusiasts, professionals, and anyone curious about the transformative power of drone mapping.